P9-BJO-793

ATLAS OF BRITISH HISTORY

OTHER BOOKS BY MARTIN GILBERT

The Churchill biography
Volume III, 'The Challenge of War', 1914–1916
Volume III, Documents, Parts I and II
Volume IV, 'The Stricken World', 1917–1922
Volume IV, Documents, Parts I, II and III
Volume V, 'The Prophet of Truth', 1922–1939
The Exchequer Years, Documents, 1922–1929
The Wilderness Years, Documents, 1929–1935
The Coming of War, Documents, 1936–1939
Volume VI, 'Finest Hour', 1939–1941
At the Admiralty, Documents, 1939–1940
Volume VII, 'Road to Victory', 1941–1945
Volume VIII, 'Never Despair', 1945–1965

Historical works
The Appeasers (with Richard Gott)
The European Powers, 1900–1945
The Roots of Appeasement
Britain and Germany Between the Wars (documents)
Plough My Own Furrow: the Life of Lord Allen of Hurtwood (documents)
Servant of India: Diaries of the Viceroy's Private Secretary (documents)
Sir Horace Rumbold: Portrait of a Diplomat
Churchill: a Photographic Portrait
Churchill's Political Philosophy
Auschwitz and the Allies
Exile and Return: the Struggle for Jewish Statehood
The Jews of Hope: the Plight of Soviet Jewry Today
Shcharansky: Hero of our Time
Jerusalem: Rebirth of a City, 1838–1898
Final Journey: the Fate of the Jews in Nazi Europe
The Holocaust: the Jewish Tragedy
The Second World War
Churchill, A Life

Atlases
Dent Atlas of American History
Dent Atlas of the Arab–Israeli Conflict
Dent Atlas of the First World War
Dent Atlas of the Holocaust
Dent Atlas of Jewish History
Dent Atlas of Recent History (*in preparation*)
Dent Atlas of Russian History
The Jews of Arab Lands: Their History in Maps
The Jews of Russia: Their History in Maps
Jerusalem: Illustrated History Atlas
Children's Illustrated Bible Atlas

ATLAS OF BRITISH HISTORY

Second edition

Martin Gilbert

Fellow of Merton College, Oxford

New York

OXFORD UNIVERSITY PRESS

1993

First published in Great Britain by
The Orion Publishing Group Limited
5 Upper St. Martin's Lane, London WC2H 9EA

Published in the United States of America by
Oxford University Press, Inc.
200 Madison Avenue
New York, N.Y. 10016, U.S.A.

Library of Congress
Cataloging-in-Publication Data
Gilbert, Martin, 1936–
Atlas of British history / Martin Gilbert.
p. cm.
Rev. ed. of: British history atlas / Martin Gilbert. 1969.
Includes bibliographical references and index.
ISBN 0–19–521040–9 (hardback)
ISBN 0–19–521060–3 (paperback)
1. Great Britain—Historical geography—Maps. I. Gilbert.
Martin, 1936– British history atlas.
G1811.SIG5 1993 <G&M>
911. 42—dc20

93–21922
CIP
MAP

Printing (last digit): 9 8 7 6 5 4 3 2

Printed in Great Britain

Preface

The maps in this atlas are intended to provide a visual introduction to British history. I have used the word 'British' in its widest sense, including when relevant England, Scotland, Ireland and Wales, the changing overseas empire, the wars and treaties in which Britain engaged, the alliances in time of peace, the growth of industry and trade, and, on five of the maps, famine and plague.

The story of the British Isles forms the central theme. I have included maps to illustrate economic, social and political problems as well as territorial and military ones. I hope this atlas will help to show that there is more to British history than Hastings and Crécy, Blenheim and Waterloo, Passchendaele and Dunkirk, all of which moments of glory I have tried to put in their wider, and no less important, contexts.

For the maps covering the period before the Norman Conquest the sources are often conflicting on specific details. I have therefore drawn these maps on the basis of probability. In many instances precise knowledge of early frontiers is lacking. I have tried nevertheless to give a clear if also, of necessity, an approximate picture.

As British history advances from wattle huts to timber mansions, and thence on to steel and concrete, so too do the number and variety of facts available to the historian. This is reflected in the maps themselves. I have tried to avoid too complex or too cluttered a page; but a map cannot always satisfy all the demands made upon it, and only the reader can judge where clarity of design and sufficiency of information have been successfully combined.

I am under an obligation of gratitude to those historians and colleagues who kindly scrutinised my draft maps at an early stage, and who made many suggestions for their scope and improvement; in particular Dr J. M. Wallace-Hadrill, Dr Roger Highfield, Mr Ralph Davis, Mr T. F. R. G. Braun, Dr C. C. Davies and Miss Barbara Malament. When the maps were more completed, they were checked by Mr Adrian Scheps, Mr Edmund Ranallo, Mrs Elizabeth Goold, Mr Tony Lawdham and Mrs Jean Kelly, to all of whom my thanks are due.

Twenty-five years have passed since the first edition of this atlas. Within a year of its publication, violence in Northern Ireland re-emerged at the centre of the political stage: I have drawn three new maps to reflect this. The evolution of the European Community has led to growing British participation in Europe, culminating in the Maastricht Treaty of February 1992 and the Edinburgh Summit of December 1992, both of which are a part of the new maps. The Falkland Islands and Persian Gulf wars are included, as are the natural and man-made disasters of the past forty years. Also mapped are many of the problems and challenges of the 1990s, among them asylum, charity, homelessness, unemployment, trade, education, religious diversity, and ethnic minorities. Britain's oil and gas resources are a new feature, as is the most recent phase of the reduction of British overseas possessions, her dwindling military and naval commitments world wide, and her new overseas responsibilities.

The first 118 maps were produced for this atlas by Arthur Banks and his team of cartographers, including Terry Bicknell. The new maps in this edition were produced by Tim Aspden and Robert Bradbrook; I have been helped considerably in the task of compiling them by Abe Eisenstat and Kay Thomson. For their help in providing material for this volume, I would also like to thank the Information Officer, Private Secretary's Office, Buckingham Palace; the Board of Deputies of British Jews, Central Information Desk; the Building Societies Association Press Office; the Lesotho High Commission; the Race Relations Commission; the Refugee Arrivals Project, London Airport; and the Royal Ulster Constabulary Press Office, Belfast.

24 June 1993

MARTIN GILBERT
Merton College, Oxford

Maps

THE CELTS IN BRITAIN BY 50 BC
PICTI
OTADINI
SELGOVAE
Tweed
NOVANTAE
BRIGANTES
Ouse
Trent
DEGEANGLI
CORNOVII
ICENI
ORDOVICES
CORITANI
Welland
Severn
TRINOVANTES
DEMETAE
SILURES
CATUVELLAUNI
DOBUNI
ATREBATES
Thames
CANTII
BELGAE
REGNI
DUMNONII
DUROTRIGES
0
50
Miles
Celtic Tribes

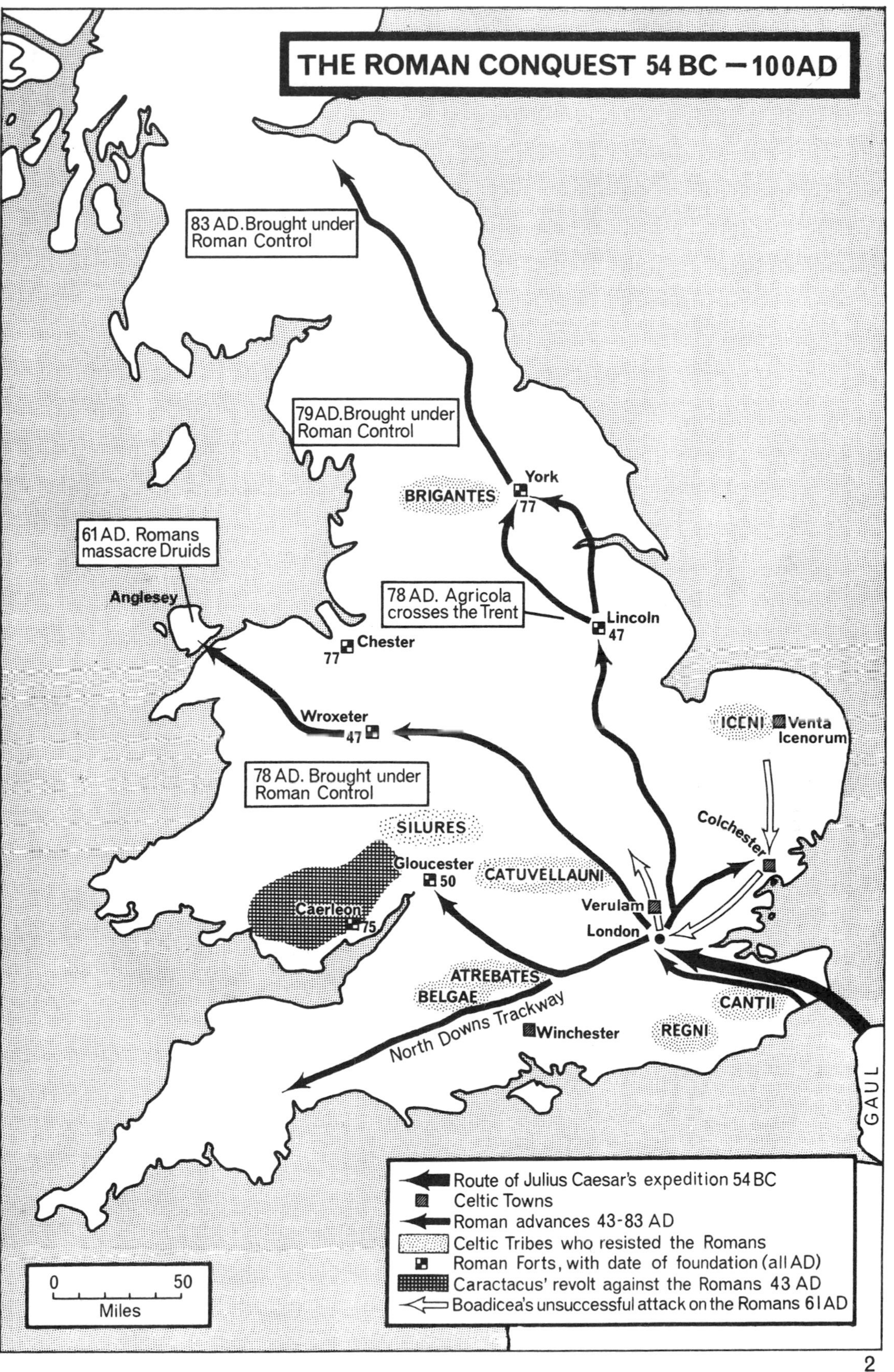

THE ROMAN CONQUEST 54 BC — 100AD
83 AD. Brought under Roman Control
79 AD. Brought under Roman Control
61 AD. Romans massacre Druids
Anglesey
York
77
BRIGANTES
78 AD. Agricola crosses the Trent
Lincoln
47
Chester
77
Wroxeter
47
ICENI
Venta Icenorum
78 AD. Brought under Roman Control
SILURES
Gloucester
50
CATUVELLAUNI
Colchester
Caerleon
75
Verulam
London
ATREBATES
BELGAE
North Downs Trackway
Winchester
REGNI
CANTII
GAUL
Route of Julius Caesar's expedition 54 BC
Celtic Towns
Roman advances 43-83 AD
Celtic Tribes who resisted the Romans
Roman Forts, with date of foundation (all AD)
Caractacus' revolt against the Romans 43 AD
Boadicea's unsuccessful attack on the Romans 61 AD
0
50
Miles

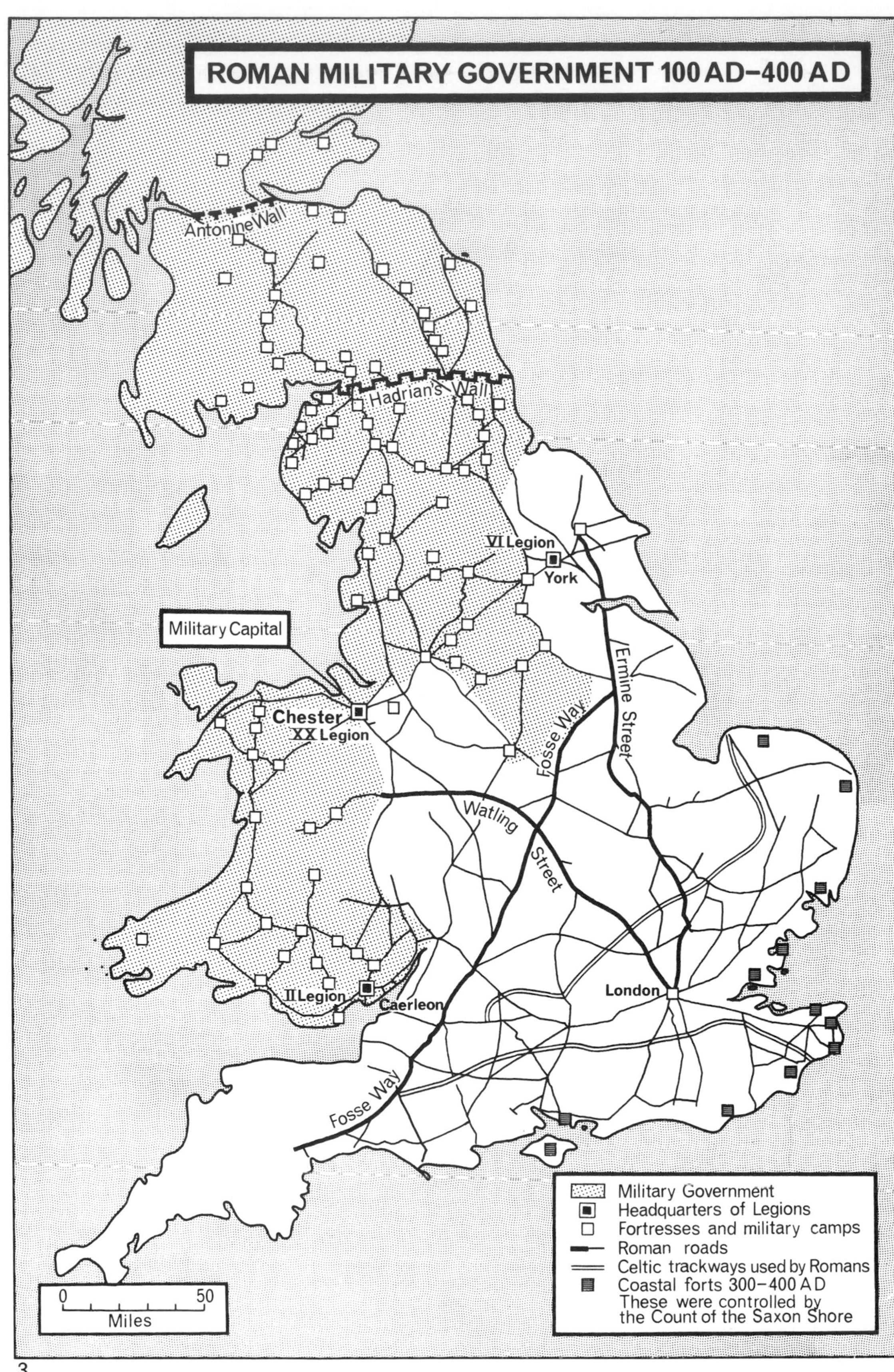
ROMAN MILITARY GOVERNMENT 100 AD–400 AD
Antonine Wall
Hadrian's Wall
VI Legion
York
Military Capital
Chester
XX Legion
Ermine Street
Fosse Way
Watling Street
II Legion
Caerleon
London
Fosse Way
Military Government
Headquarters of Legions
Fortresses and military camps
Roman roads
Celtic trackways used by Romans
Coastal forts 300–400 AD
These were controlled by the Count of the Saxon Shore
0
50
Miles

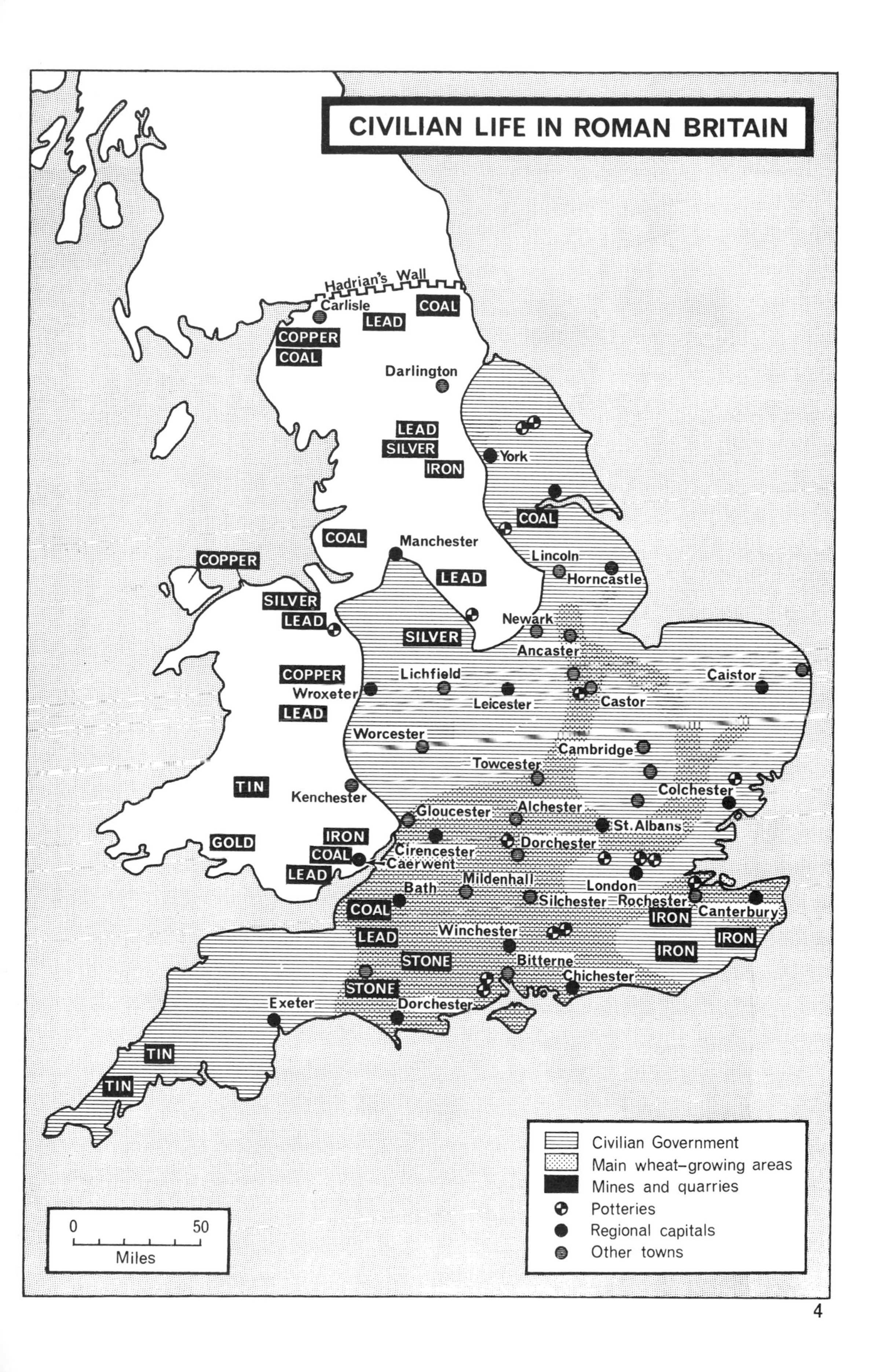
CIVILIAN LIFE IN ROMAN BRITAIN
Hadrian's Wall
Carlisle
COAL
LEAD
COPPER
COAL
Darlington
LEAD
SILVER
IRON
York
COAL
COAL
Manchester
COPPER
Lincoln
Horncastle
LEAD
SILVER
LEAD
Newark
SILVER
Ancaster
COPPER
Lichfield
Caistor
Wroxeter
Leicester
Castor
LEAD
Worcester
Cambridge
Towcester
TIN
Kenchester
Colchester
Gloucester
Alchester
St. Albans
GOLD
IRON
Dorchester
COAL
Cirencester
Caerwent
LEAD
Mildenhall
London
Bath
Silchester
Rochester
COAL
IRON
Canterbury
Winchester
LEAD
IRON
IRON
STONE
Bitterne
STONE
Chichester
Exeter
Dorchester
TIN
TIN
Civilian Government
Main wheat-growing areas
Mines and quarries
Potteries
Regional capitals
Other towns
0
50
Miles

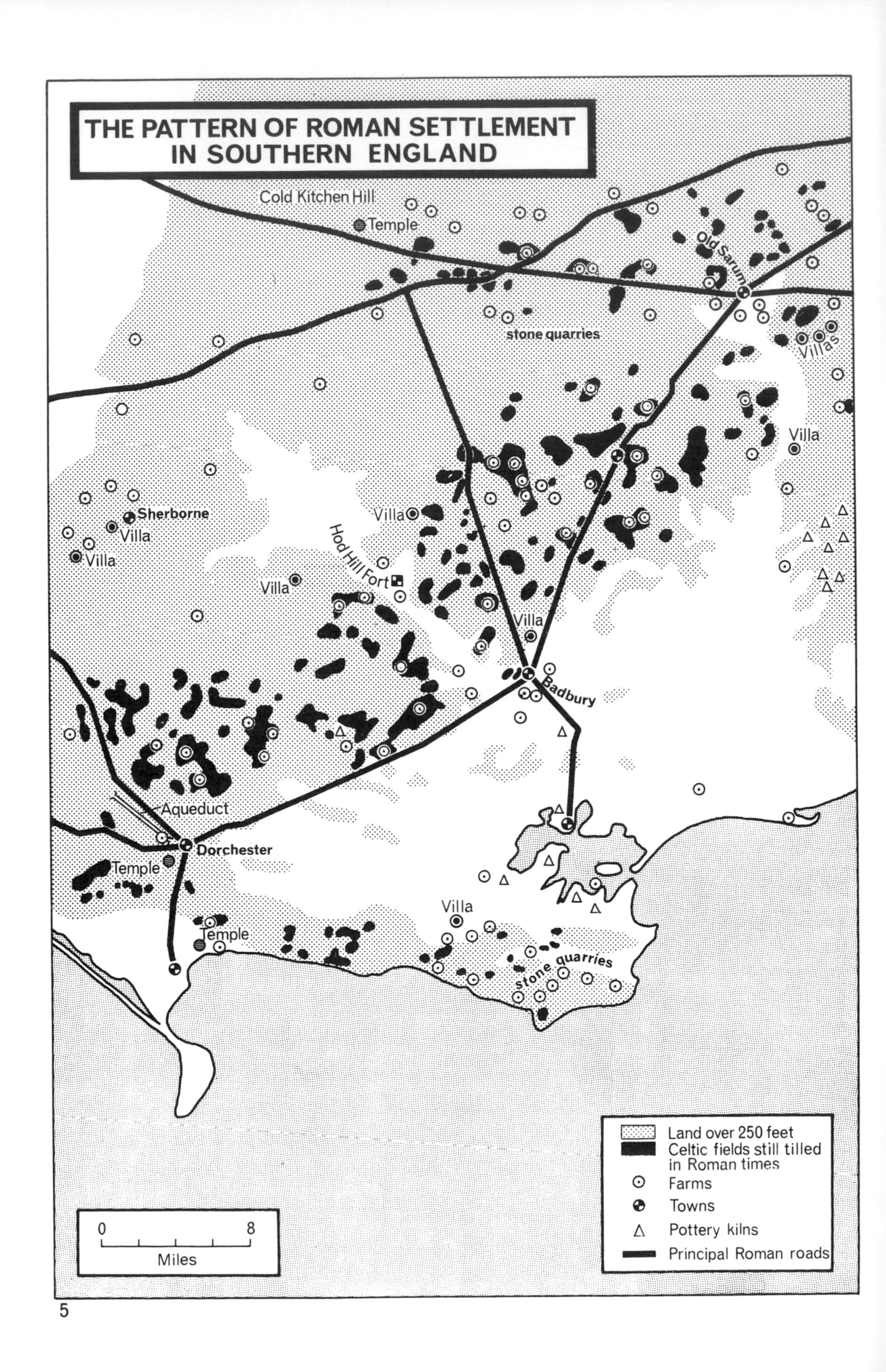
THE PATTERN OF ROMAN SETTLEMENT IN SOUTHERN ENGLAND
Cold Kitchen Hill
Temple
Old Sarum
stone quarries
Villas
Villa
Sherborne
Villa
Villa
Villa
Hod Hill Fort
Villa
Villa
Badbury
Aqueduct
Dorchester
Temple
Temple
Villa
stone quarries
Land over 250 feet
Celtic fields still tilled in Roman times
Farms
Towns
Pottery kilns
Principal Roman roads
0
8
Miles

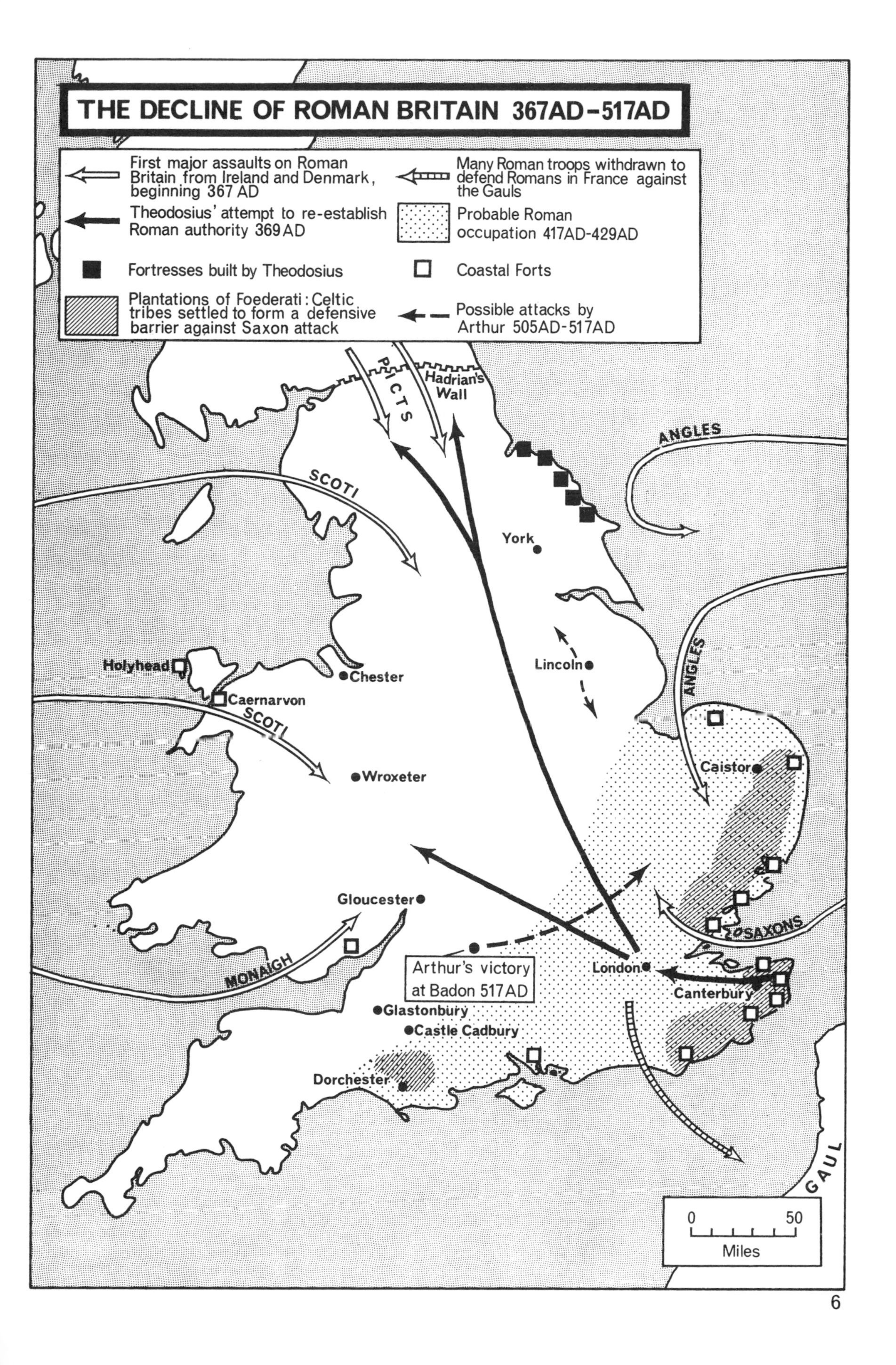
THE DECLINE OF ROMAN BRITAIN 367AD-517AD
First major assaults on Roman Britain from Ireland and Denmark, beginning 367 AD
Many Roman troops withdrawn to defend Romans in France against the Gauls
Theodosius' attempt to re-establish Roman authority 369 AD
Probable Roman occupation 417AD-429AD
Fortresses built by Theodosius
Coastal Forts
Plantations of Foederati: Celtic tribes settled to form a defensive barrier against Saxon attack
Possible attacks by Arthur 505AD-517AD
PICTS
Hadrian's Wall
ANGLES
SCOTI
York
Lincoln
Holyhead
Chester
Caernarvon
SCOTI
ANGLES
Caistor
Wroxeter
Gloucester
SAXONS
MONAIGH
Arthur's victory at Badon 517AD
London
Canterbury
Glastonbury
Castle Cadbury
Dorchester
GAUL
0
50
Miles

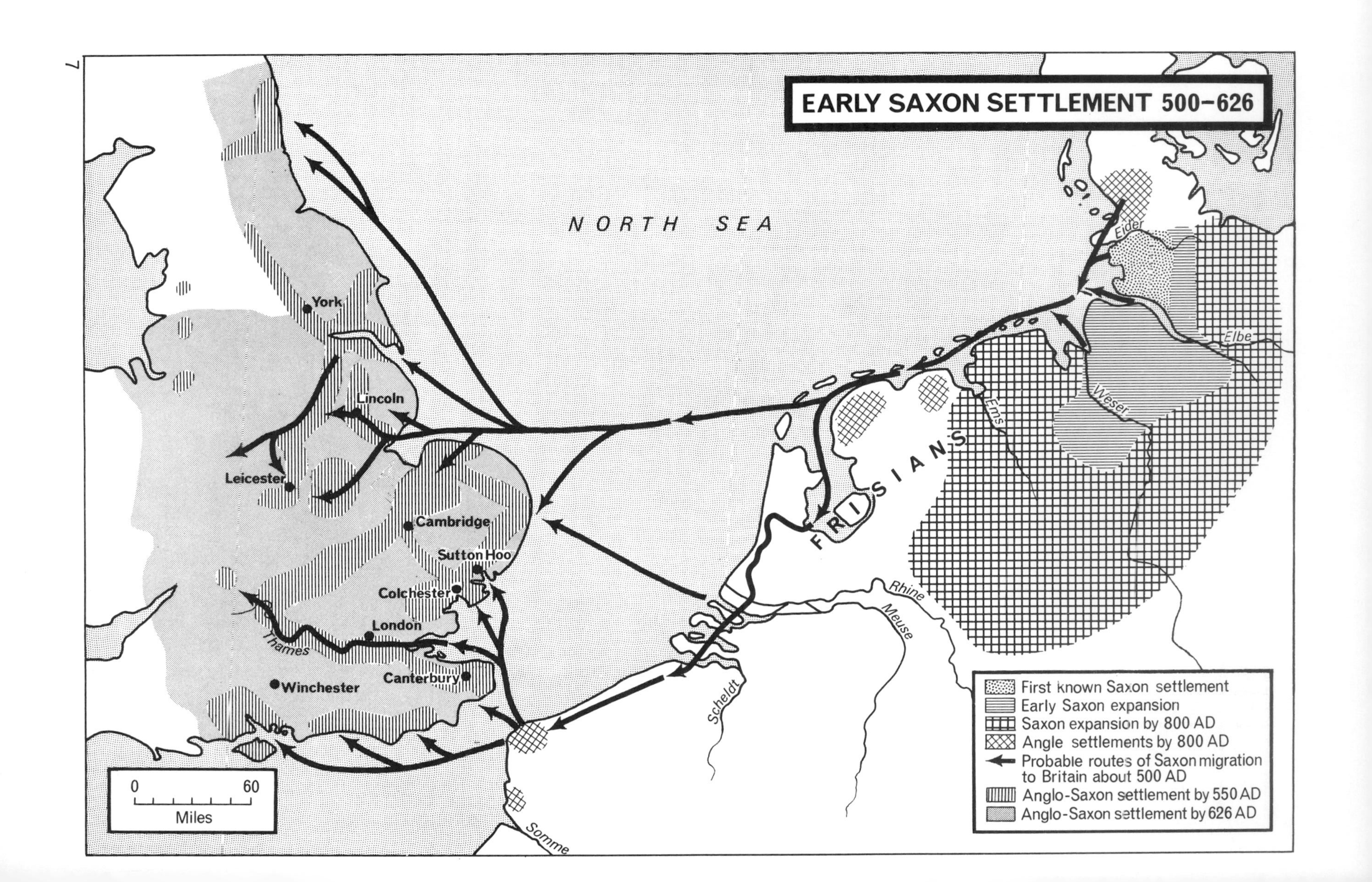
EARLY SAXON SETTLEMENT 500–626
NORTH SEA
FRISIANS
York
Lincoln
Leicester
Cambridge
Sutton Hoo
Colchester
London
Thames
Canterbury
Winchester
Eider
Elbe
Weser
Ems
Rhine
Meuse
Scheldt
Somme
0
60
Miles
First known Saxon settlement
Early Saxon expansion
Saxon expansion by 800 AD
Angle settlements by 800 AD
Probable routes of Saxon migration to Britain about 500 AD
Anglo-Saxon settlement by 550 AD
Anglo-Saxon settlement by 626 AD

SAXON KINGDOMS AND BRETWALDASHIPS 630-829

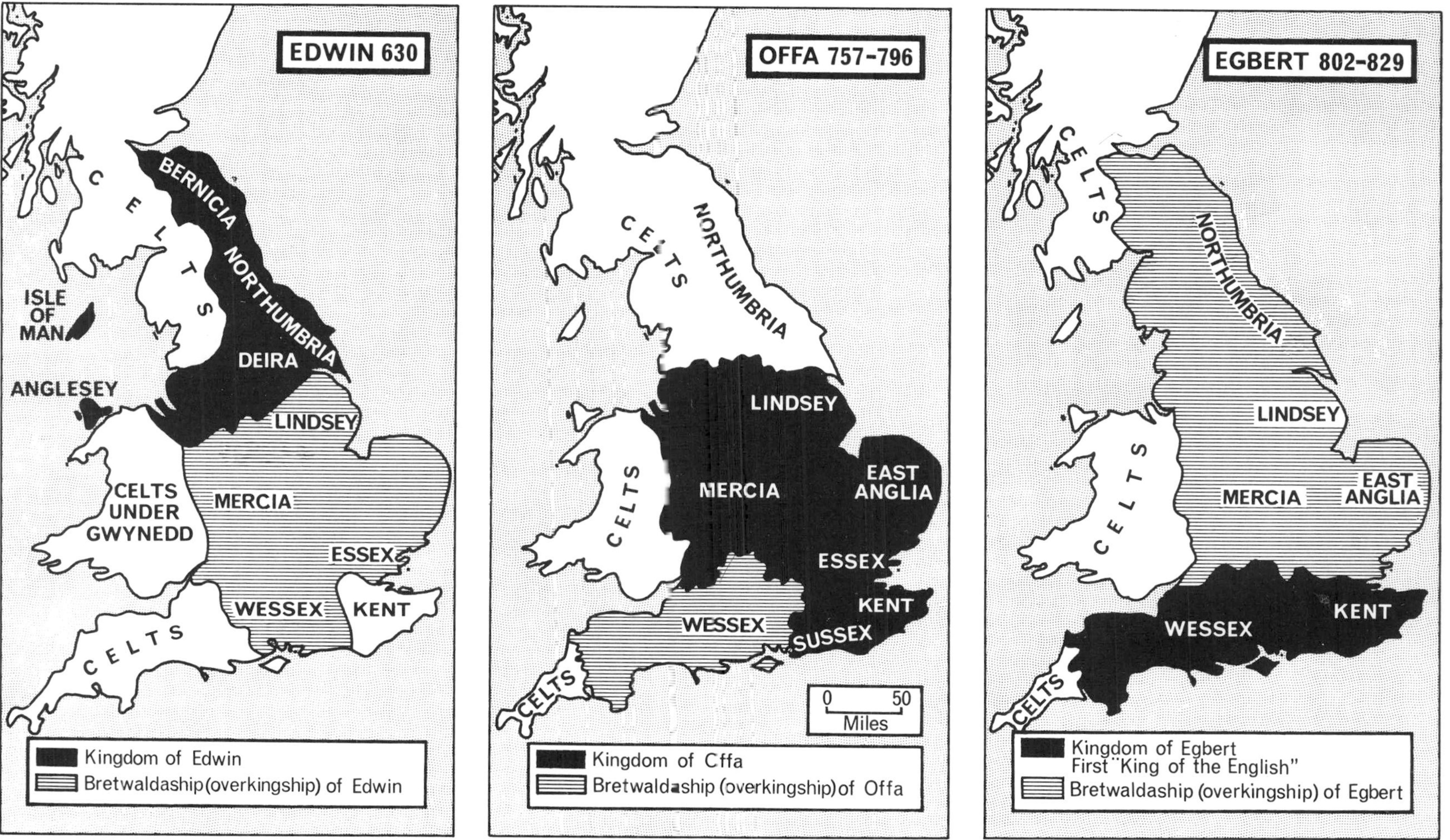

THE CHURCH 700-850
†Abercorn
Coldingham
Lindisfarne
†Melrose
LINDISFARNE
†Coquet Island
WHITHORN
Whithorn
Hexham
HEXHAM
†Tynemouth
†Jarrow
†Monkwearmouth
Hartlepool
†Gainford
Gilling
†Sockburn
Whitby
Lastingham
Hackness
YORK
†Ripon
York
Barrow
Syddensis Civitas
(site not known)
LINDSEY
LICHFIELD
Repton
†Breedon
Lichfield
Elmham
ELMHAM
Peterborough
Leicester
†Oundle
Ely
Dunwich
†Bury St. Edmunds
†Brixworth
LEICESTER
DUNWICH
HEREFORD
Worcester
Hereford
WORCESTER
DORCHESTER
LONDON
Malmesbury
Abingdon
Dorchester
Barking
London
Reculver
Minster
ROCHESTER
Canterbury
Dover
Folkestone
Lyminge
CANTERBURY
Woking
WINCHESTER
Glastonbury
Winchester
Sherborne
†Tisbury
Nursling
SELSEY
Selsey
SHERBORNE
Wimborne
Exeter
† Religious houses founded by 850
Double houses where monks and nuns lived under the rule of an abbess
Approximate diocesan boundaries
Diocesan seats
Archbishoprics
0
50
Miles

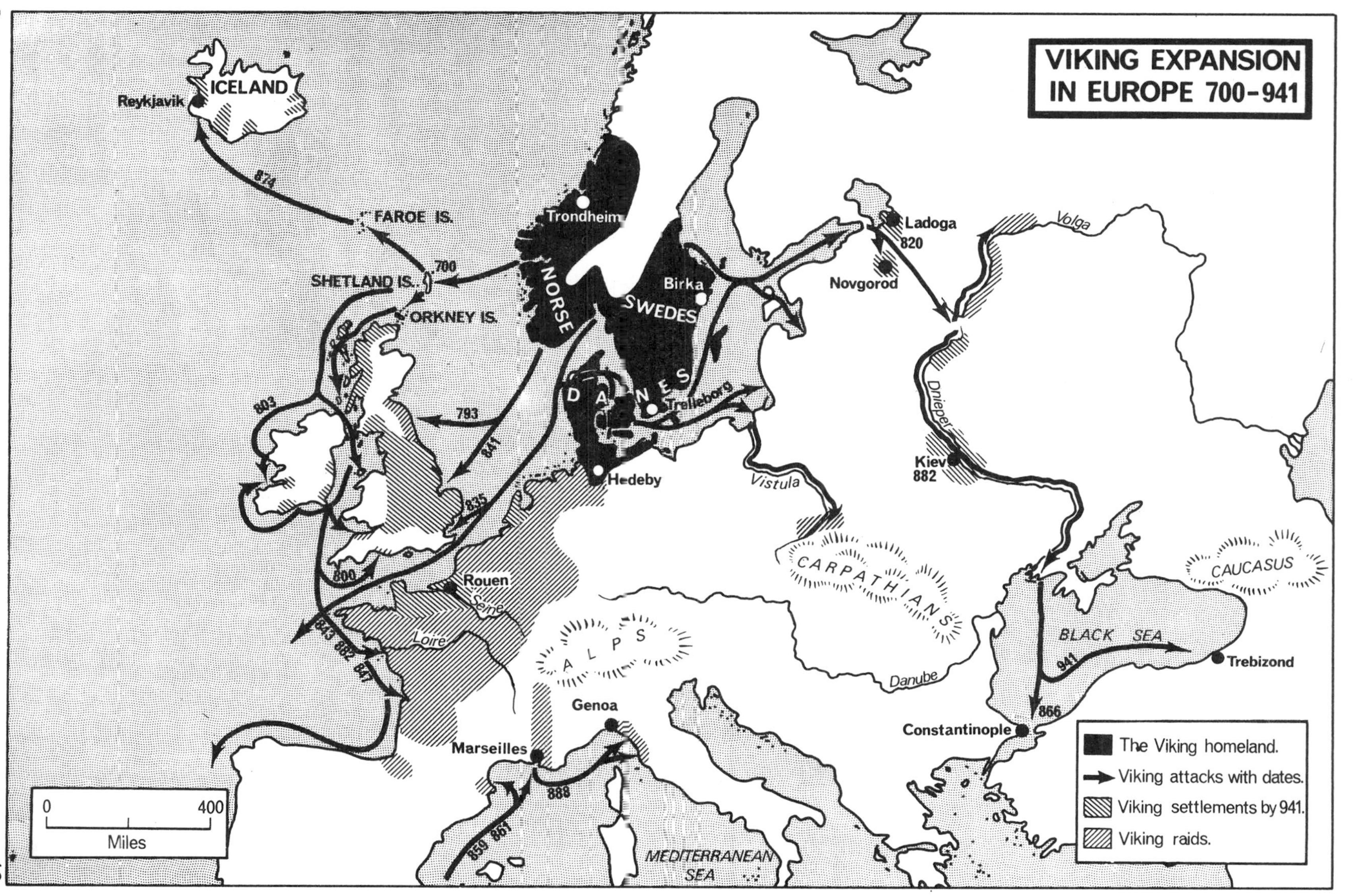
VIKING EXPANSION
IN EUROPE 700-941
The Viking homeland.
Viking attacks with dates.
Viking settlements by 941.
Viking raids.
ICELAND
Reykjavik
874
FAROE IS.
700
SHETLAND IS.
ORKNEY IS.
Trondheim
NORSE
Birka
SWEDES
DANES
Trelleborg
Hedeby
803
793
841
835
800
843 882 847
Rouen
Seine
Loire
Ladoga
820
Novgorod
Volga
Dnieper
Kiev
882
Vistula
CARPATHIANS
CAUCASUS
BLACK SEA
941
866
Trebizond
Constantinople
Danube
ALPS
Genoa
Marseilles
888
859 861
MEDITERRANEAN
SEA
0
400
Miles

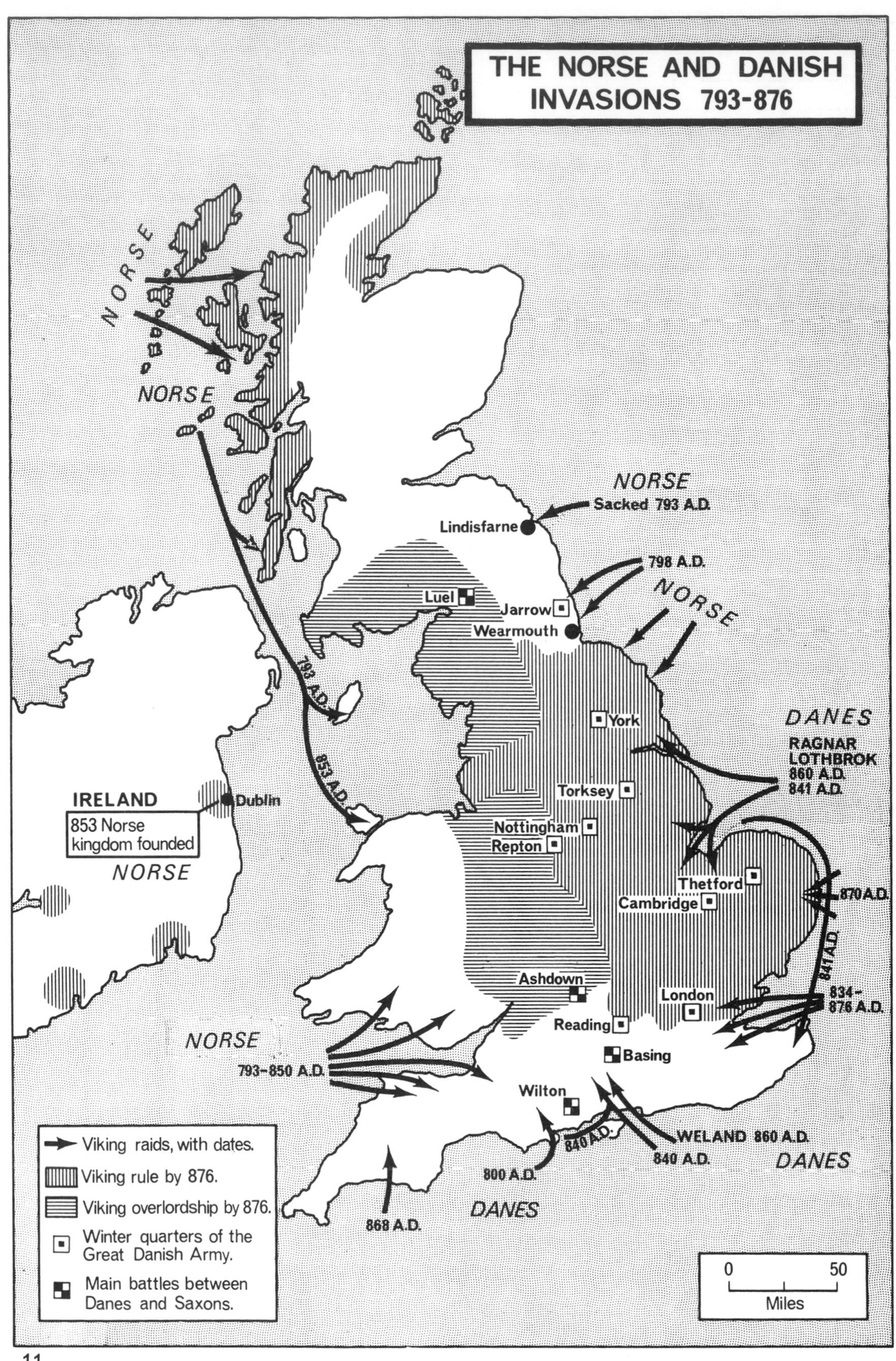
THE NORSE AND DANISH INVASIONS 793-876
NORSE
NORSE
NORSE
Sacked 793 A.D.
Lindisfarne
798 A.D.
NORSE
Luel
Jarrow
Wearmouth
793 A.D.
853 A.D.
York
DANES
RAGNAR LOTHBROK
860 A.D.
841 A.D.
IRELAND
Dublin
853 Norse kingdom founded
NORSE
Torksey
Nottingham
Repton
Thetford
Cambridge
870 A.D.
847 A.D.
Ashdown
London
834-876 A.D.
Reading
NORSE
793-850 A.D.
Basing
Wilton
840 A.D.
WELAND 860 A.D.
840 A.D.
DANES
800 A.D.
868 A.D.
DANES
Viking raids, with dates.
Viking rule by 876.
Viking overlordship by 876.
Winter quarters of the Great Danish Army.
Main battles between Danes and Saxons.
0
50
Miles

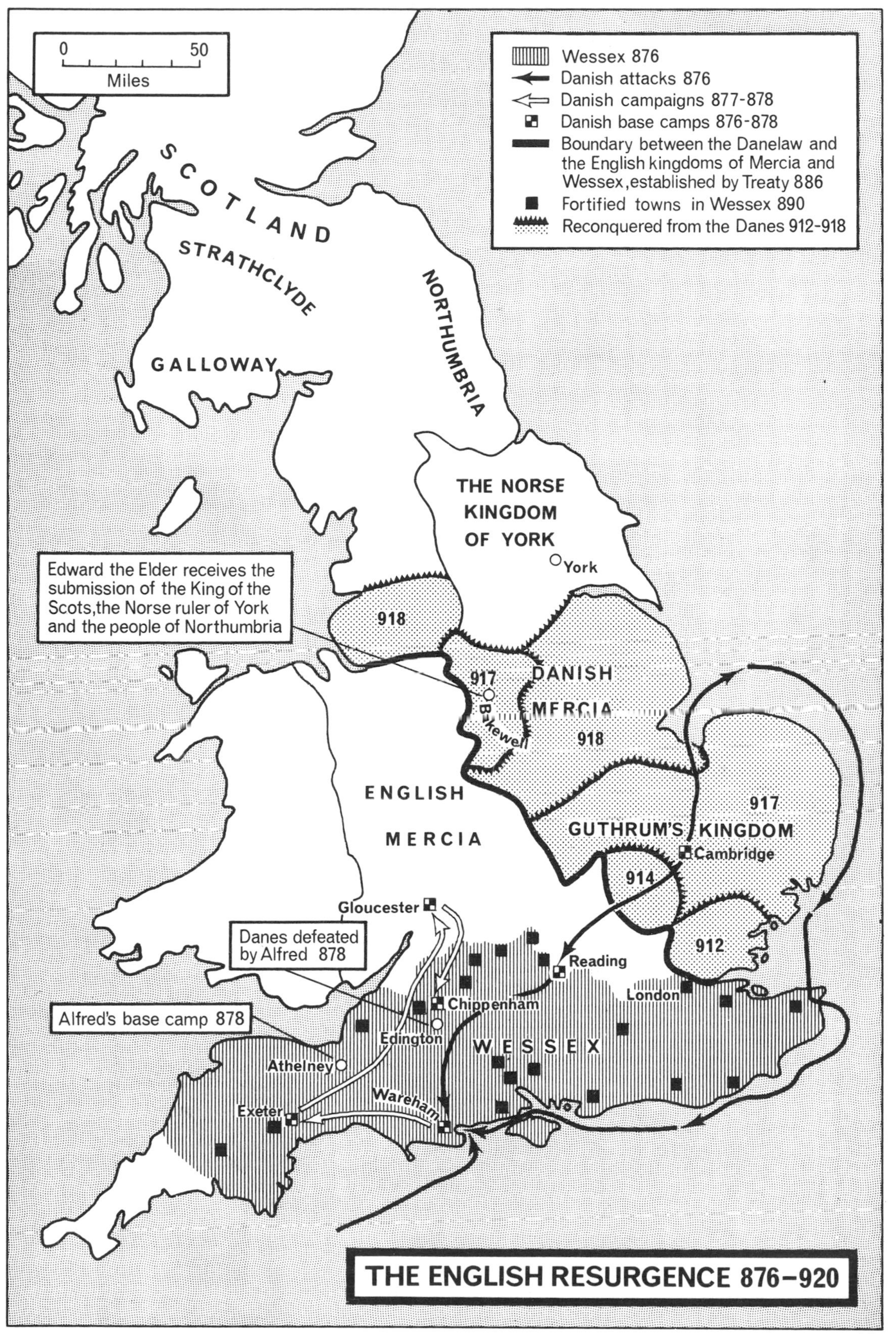

0
50
Miles
Wessex 876
Danish attacks 876
Danish campaigns 877-878
Danish base camps 876-878
Boundary between the Danelaw and the English kingdoms of Mercia and Wessex, established by Treaty 886
Fortified towns in Wessex 890
Reconquered from the Danes 912-918
SCOTLAND
STRATHCLYDE
GALLOWAY
NORTHUMBRIA
THE NORSE KINGDOM OF YORK
York
Edward the Elder receives the submission of the King of the Scots, the Norse ruler of York and the people of Northumbria
918
917
DANISH MERCIA
Bakewell
918
ENGLISH MERCIA
917
GUTHRUM'S KINGDOM
Cambridge
914
912
Gloucester
Danes defeated by Alfred 878
Reading
Chippenham
London
Alfred's base camp 878
Edington
WESSEX
Athelney
Wareham
Exeter
THE ENGLISH RESURGENCE 876–920

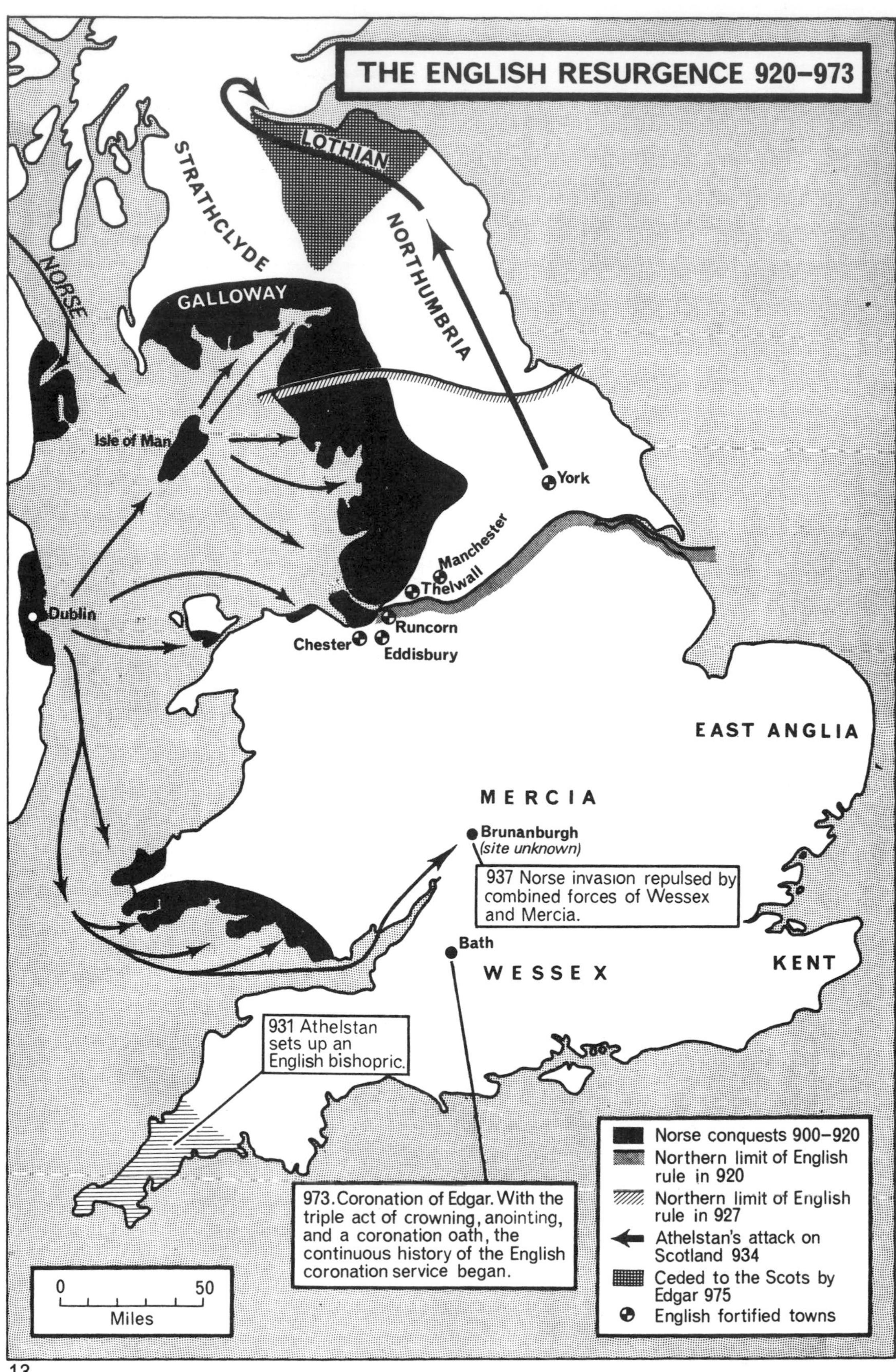
THE ENGLISH RESURGENCE 920–973
LOTHIAN
STRATHCLYDE
NORTHUMBRIA
NORSE
GALLOWAY
Isle of Man
York
Manchester
Thelwall
Dublin
Runcorn
Chester
Eddisbury
EAST ANGLIA
MERCIA
Brunanburgh
(site unknown)
937 Norse invasion repulsed by combined forces of Wessex and Mercia.
Bath
WESSEX
KENT
931 Athelstan sets up an English bishopric.
973. Coronation of Edgar. With the triple act of crowning, anointing, and a coronation oath, the continuous history of the English coronation service began.
0
50
Miles
Norse conquests 900–920
Northern limit of English rule in 920
Northern limit of English rule in 927
Athelstan's attack on Scotland 934
Ceded to the Scots by Edgar 975
English fortified towns

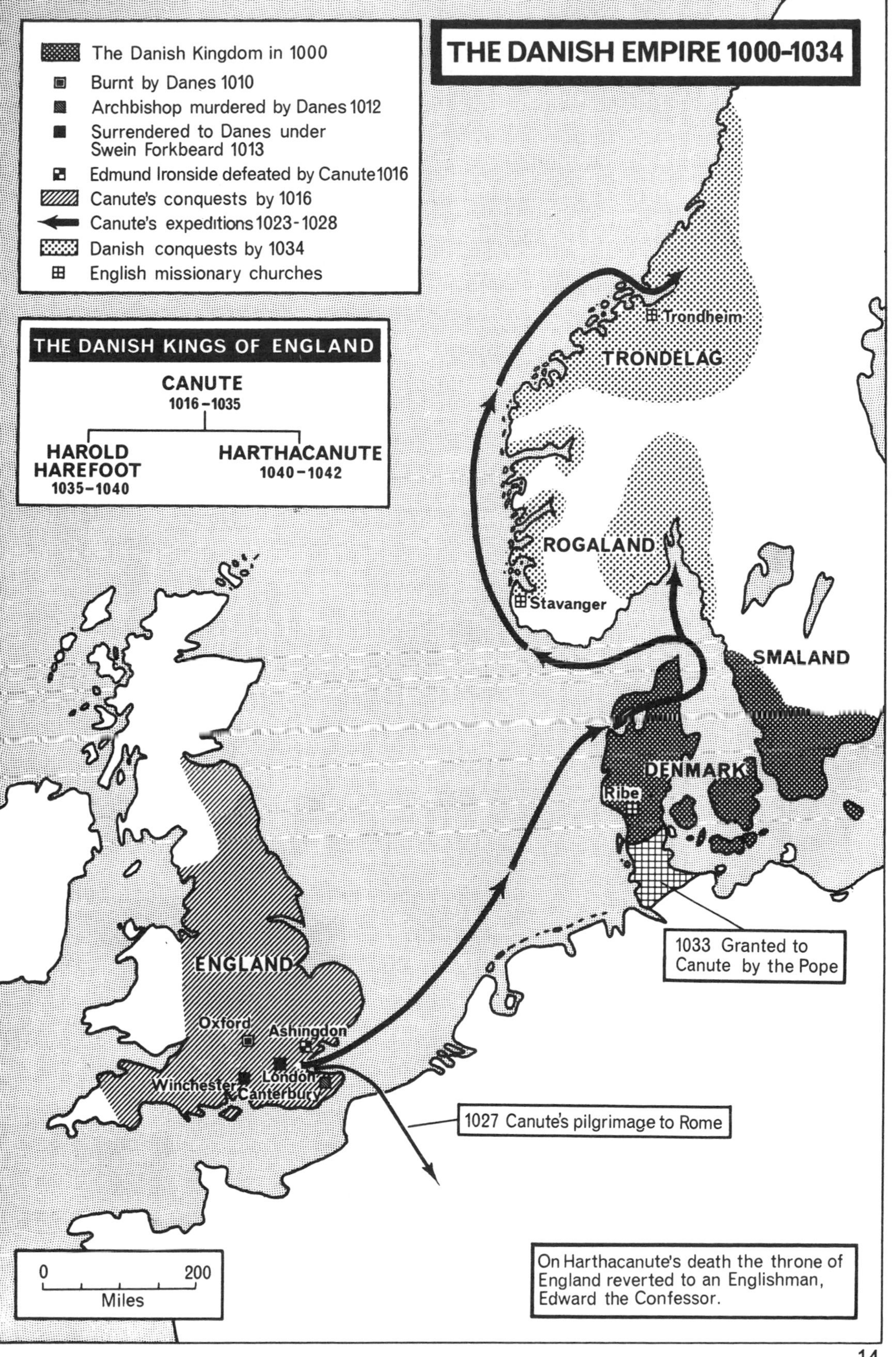

THE DANISH EMPIRE 1000-1034
The Danish Kingdom in 1000
Burnt by Danes 1010
Archbishop murdered by Danes 1012
Surrendered to Danes under Swein Forkbeard 1013
Edmund Ironside defeated by Canute 1016
Canute's conquests by 1016
Canute's expeditions 1023-1028
Danish conquests by 1034
English missionary churches
THE DANISH KINGS OF ENGLAND
CANUTE
1016-1035
HAROLD HAREFOOT
1035-1040
HARTHACANUTE
1040-1042
Trondheim
TRONDELAG
ROGALAND
Stavanger
SMALAND
DENMARK
Ribe
ENGLAND
Oxford
Ashingdon
London
Winchester
Canterbury
1033 Granted to Canute by the Pope
1027 Canute's pilgrimage to Rome
0
200
Miles
On Harthacanute's death the throne of England reverted to an Englishman, Edward the Confessor.

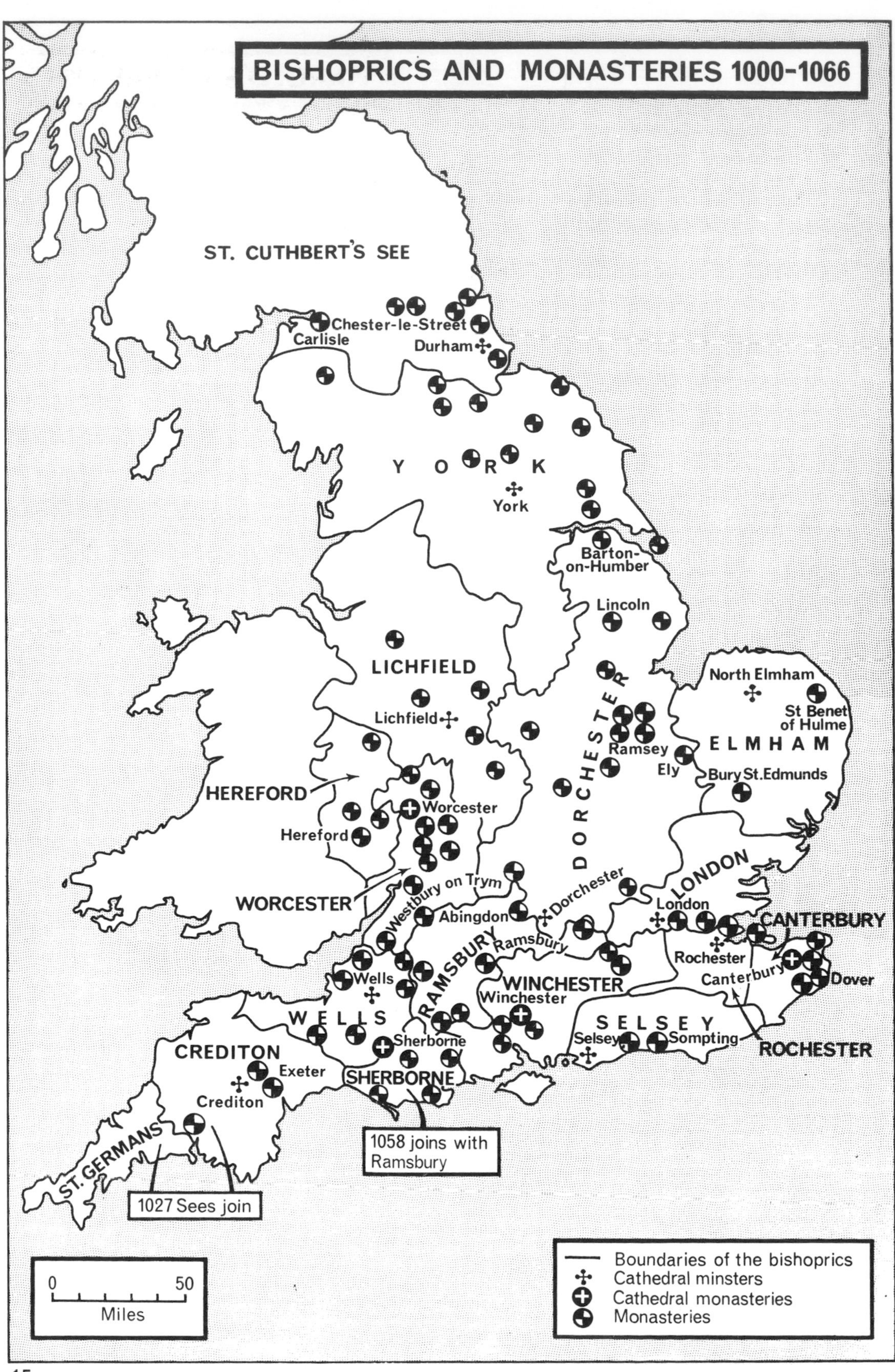
BISHOPRICS AND MONASTERIES 1000-1066
ST. CUTHBERT'S SEE
Chester-le-Street
Carlisle
Durham
YORK
York
Barton-on-Humber
Lincoln
LICHFIELD
Lichfield
DORCHESTER
North Elmham
St Benet of Hulme
ELMHAM
Ramsey
Ely
Bury St.Edmunds
HEREFORD
Worcester
Hereford
WORCESTER
Westbury on Trym
Abingdon
Dorchester
LONDON
London
CANTERBURY
Rochester
Canterbury
Dover
ROCHESTER
Ramsbury
RAMSBURY
WINCHESTER
Winchester
Wells
WELLS
SELSEY
Selsey
Sompting
Sherborne
SHERBORNE
CREDITON
Exeter
Crediton
ST. GERMANS
1058 joins with Ramsbury
1027 Sees join
0
50
Miles
Boundaries of the bishoprics
Cathedral minsters
Cathedral monasteries
Monasteries

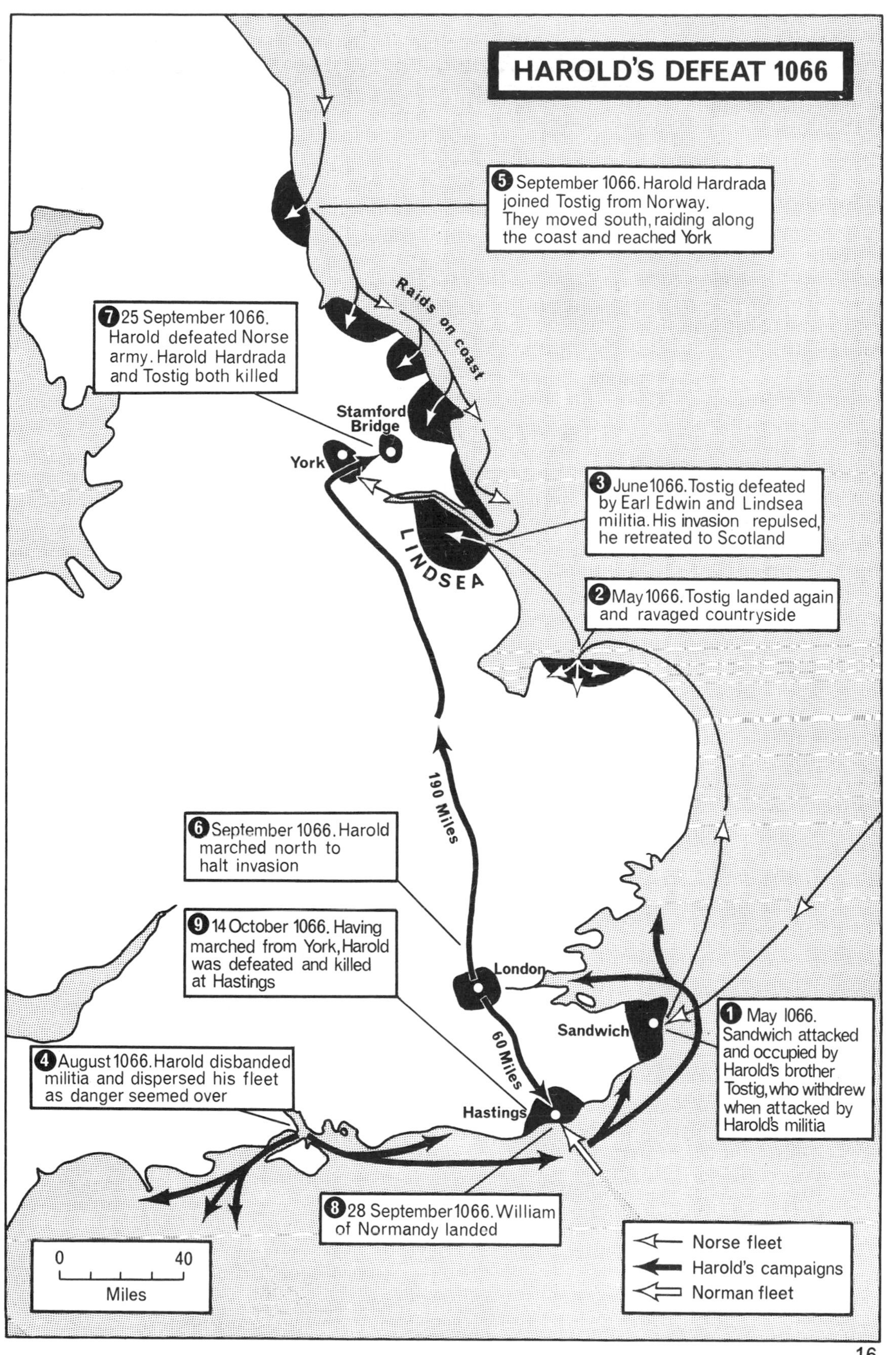
HAROLD'S DEFEAT 1066
5 September 1066. Harold Hardrada joined Tostig from Norway. They moved south, raiding along the coast and reached York
Raids on coast
7 25 September 1066. Harold defeated Norse army. Harold Hardrada and Tostig both killed
Stamford Bridge
York
3 June 1066. Tostig defeated by Earl Edwin and Lindsea militia. His invasion repulsed, he retreated to Scotland
LINDSEA
2 May 1066. Tostig landed again and ravaged countryside
190 Miles
6 September 1066. Harold marched north to halt invasion
9 14 October 1066. Having marched from York, Harold was defeated and killed at Hastings
London
Sandwich
1 May 1066. Sandwich attacked and occupied by Harold's brother Tostig, who withdrew when attacked by Harold's militia
60 Miles
4 August 1066. Harold disbanded militia and dispersed his fleet as danger seemed over
Hastings
8 28 September 1066. William of Normandy landed
0
40
Miles
Norse fleet
Harold's campaigns
Norman fleet

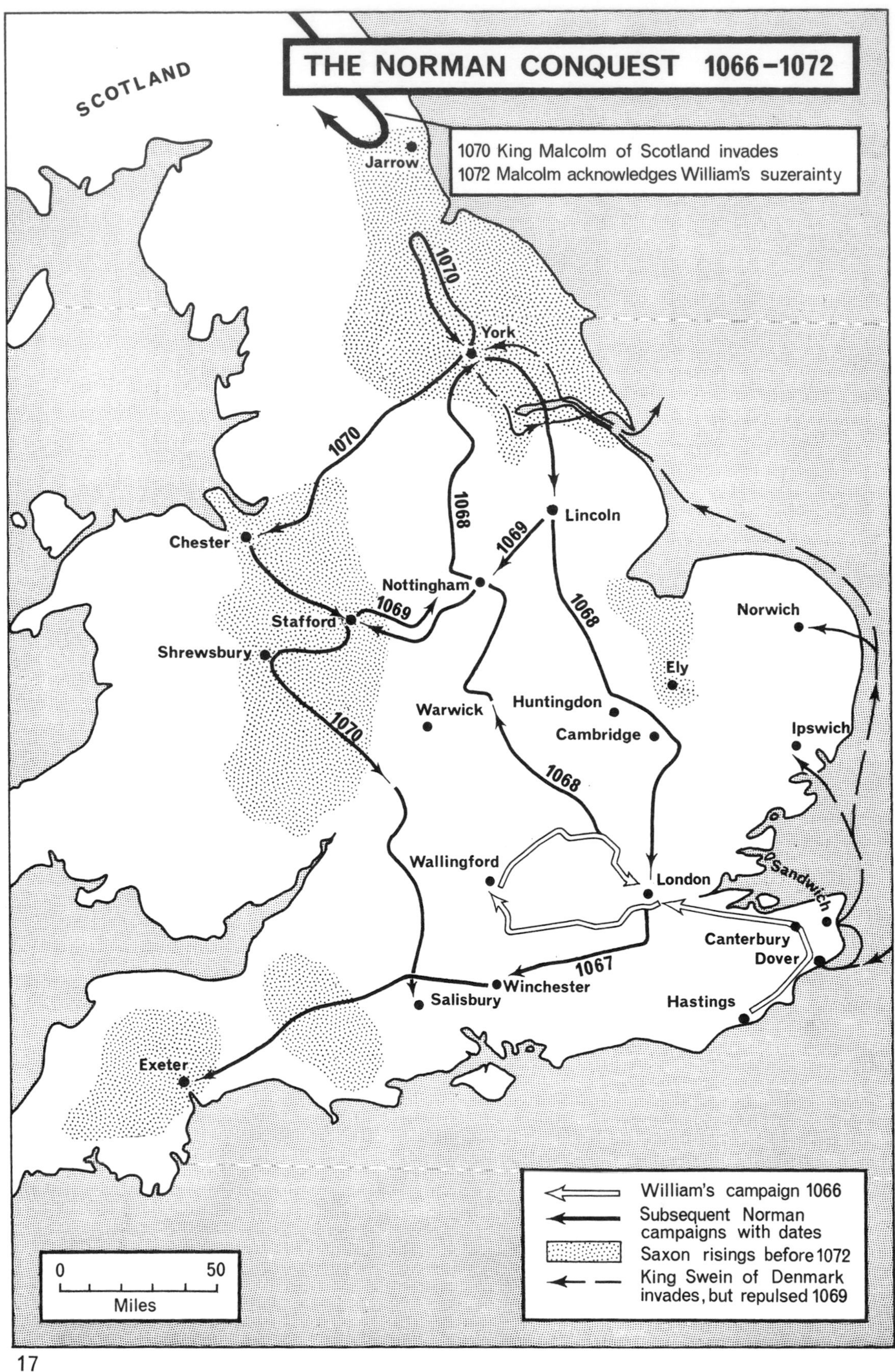
THE NORMAN CONQUEST 1066–1072
SCOTLAND
1070 King Malcolm of Scotland invades
1072 Malcolm acknowledges William's suzerainty
Jarrow
York
1070
1068
1069
1067
Lincoln
Chester
Nottingham
Stafford
Shrewsbury
Norwich
Ely
Huntingdon
Cambridge
Warwick
Ipswich
Wallingford
London
Sandwich
Canterbury
Dover
Hastings
Winchester
Salisbury
Exeter
William's campaign 1066
Subsequent Norman campaigns with dates
Saxon risings before 1072
King Swein of Denmark invades, but repulsed 1069
0
50
Miles

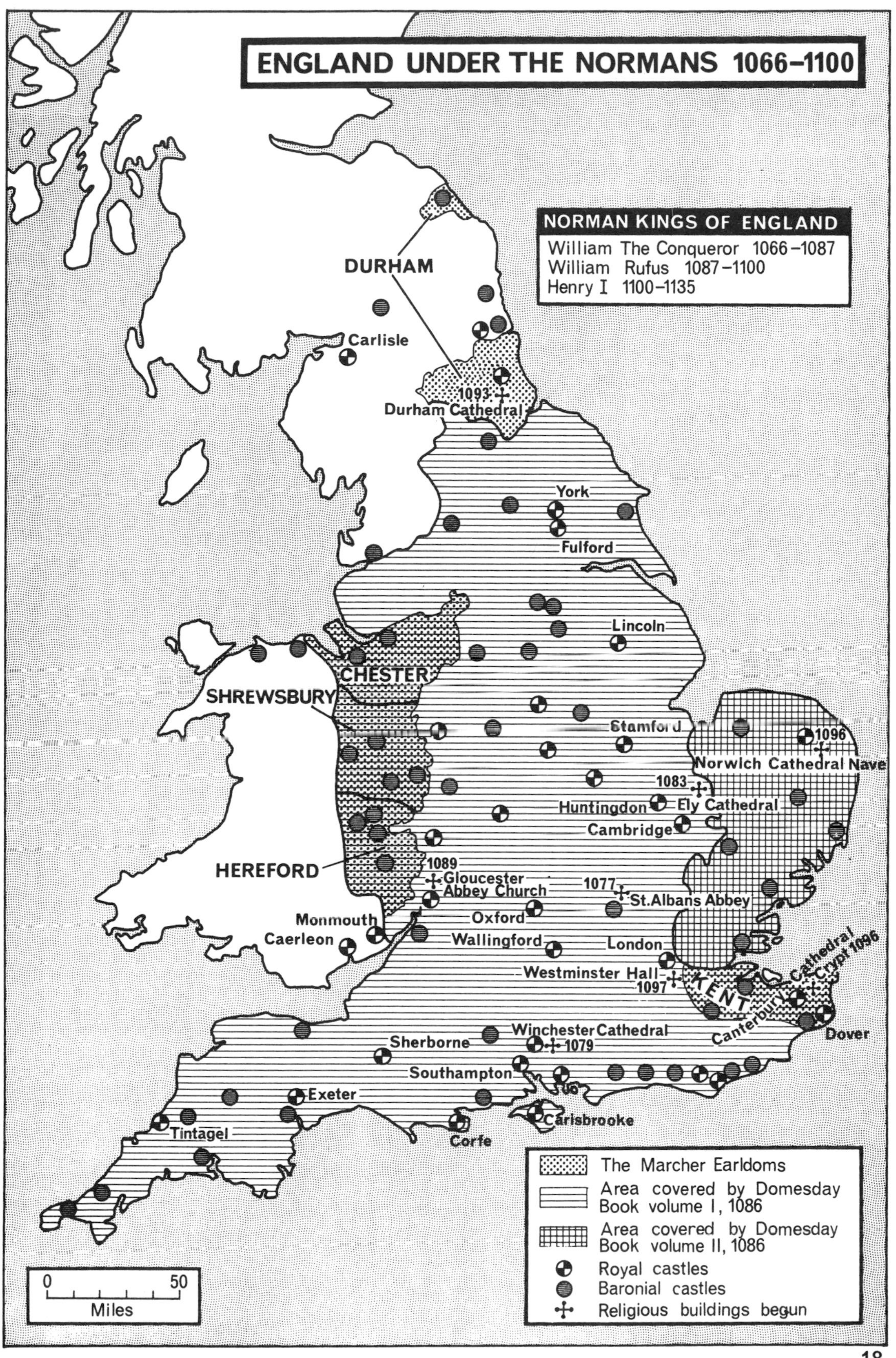

ENGLAND UNDER THE NORMANS 1066–1100
NORMAN KINGS OF ENGLAND
William The Conqueror 1066–1087
William Rufus 1087–1100
Henry I 1100–1135
DURHAM
Carlisle
1093
Durham Cathedral
York
Fulford
Lincoln
CHESTER
SHREWSBURY
Stamford
1096
Norwich Cathedral Nave
1083
Huntingdon
Ely Cathedral
Cambridge
HEREFORD
1089
Gloucester
Abbey Church
1077
St.Albans Abbey
Monmouth
Oxford
Caerleon
Wallingford
London
Westminster Hall
1097
KENT
Canterbury Cathedral Crypt 1096
Dover
Winchester Cathedral
1079
Sherborne
Southampton
Exeter
Carisbrooke
Tintagel
Corfe
The Marcher Earldoms
Area covered by Domesday Book volume I, 1086
Area covered by Domesday Book volume II, 1086
Royal castles
Baronial castles
Religious buildings begun
0
50
Miles

19

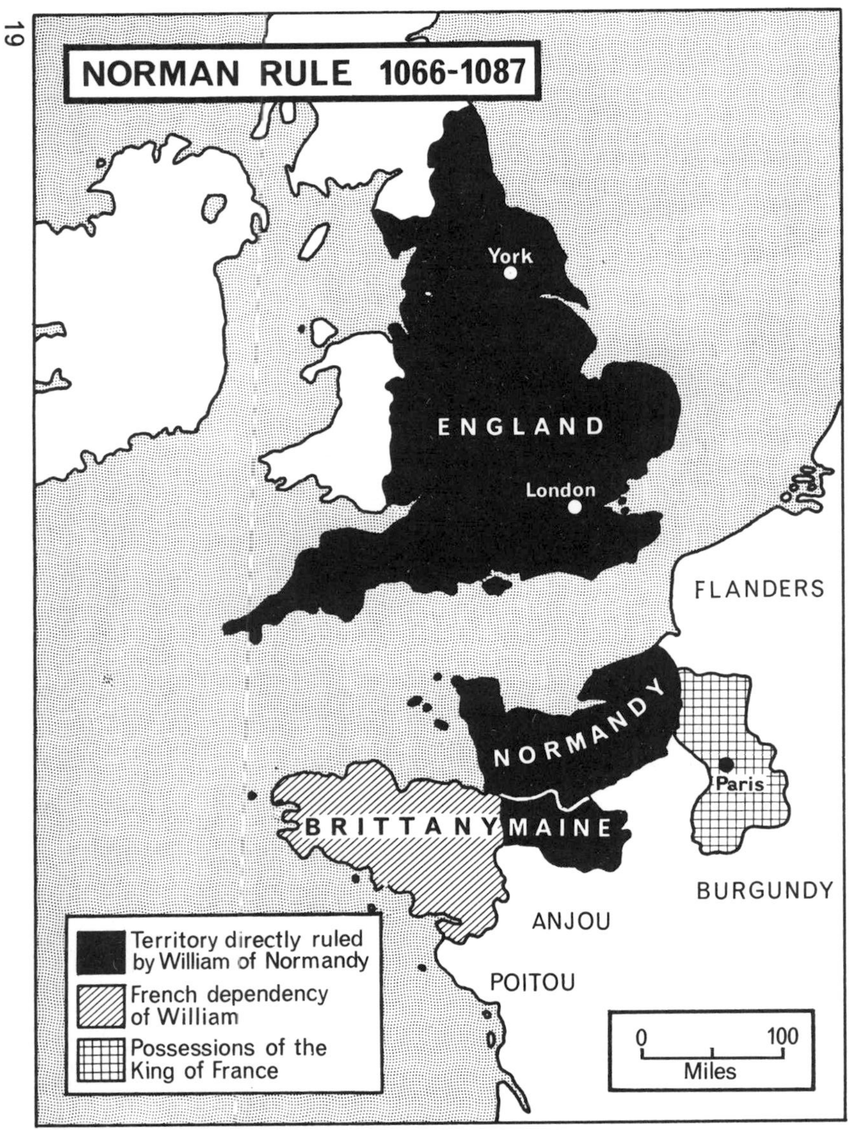

NORMAN RULE 1066-1087
York
ENGLAND
London
FLANDERS
NORMANDY
Paris
BRITTANY
MAINE
BURGUNDY
ANJOU
POITOU
Territory directly ruled by William of Normandy
French dependency of William
Possessions of the King of France
0
100
Miles

20

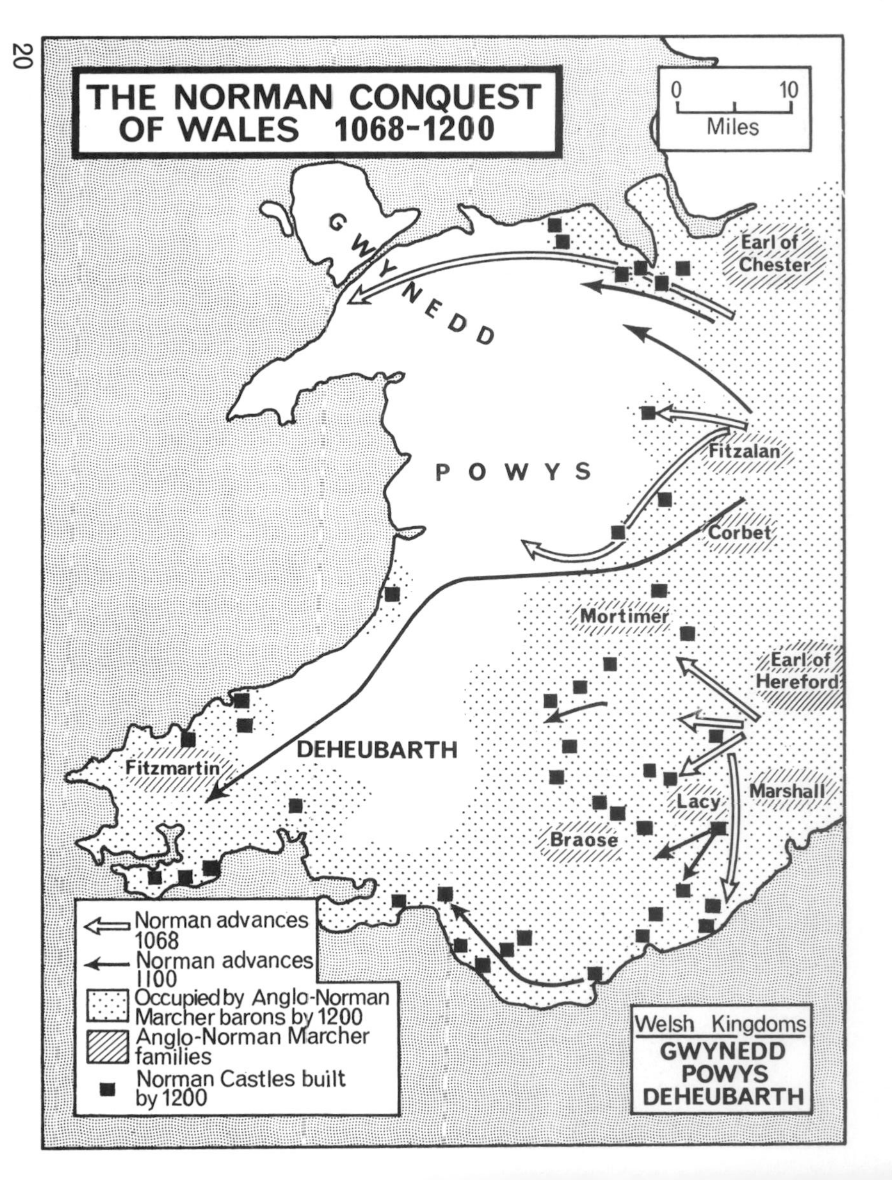

THE NORMAN CONQUEST OF WALES 1068-1200
0
10
Miles
GWYNEDD
Earl of Chester
Fitzalan
POWYS
Corbet
Mortimer
Earl of Hereford
DEHEUBARTH
Fitzmartin
Lacy
Marshall
Braose
Norman advances 1068
Norman advances 1100
Occupied by Anglo-Norman Marcher barons by 1200
Anglo-Norman Marcher families
Norman Castles built by 1200
Welsh Kingdoms
GWYNEDD
POWYS
DEHEUBARTH

IRELAND 1150
0
50
Miles
ULSTER
O'Neill
De Courcy
O'Carroll
O'Dowd
O'Rourke
O'Connor
CONNAUGHT
Annaly
MEATH
O'Kelly
O'Flaherty
De Lacy
O'Shaughnessy
Offaly
O'Dempsey
De Clare
Thomond
LEINSTER
McGillipatrick
Ormond
MacMurrough
MUNSTER
Decies
McCarthy
Desmond
O'Sullivan
Scottish settlement areas
Boundaries of the Five Kingdoms
Norman families
Irish clans

21

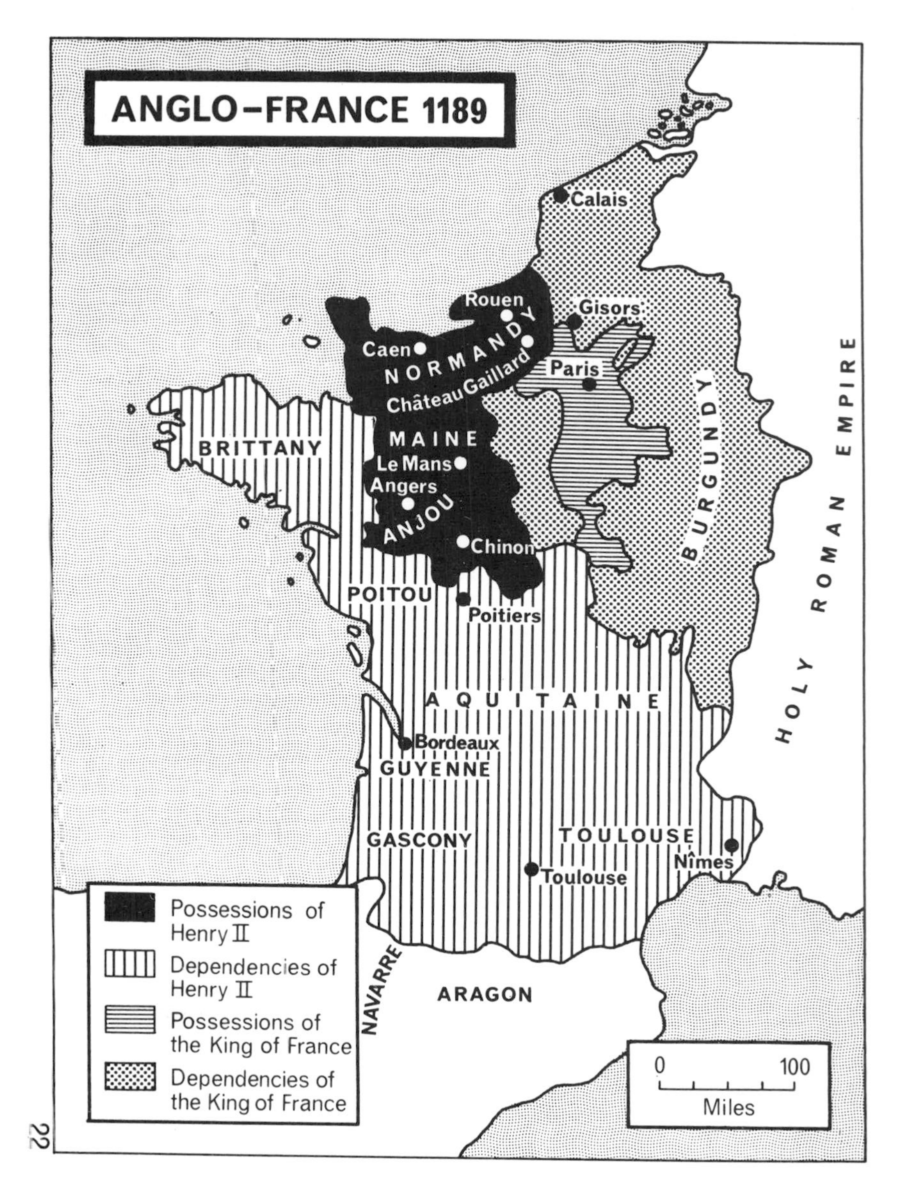
ANGLO-FRANCE 1189
Calais
Rouen
Gisors
Caen
NORMANDY
Paris
Château Gaillard
BRITTANY
MAINE
Le Mans
Angers
ANJOU
Chinon
BURGUNDY
HOLY ROMAN EMPIRE
POITOU
Poitiers
AQUITAINE
Bordeaux
GUYENNE
GASCONY
TOULOUSE
Toulouse
Nîmes
NAVARRE
ARAGON
Possessions of Henry II
Dependencies of Henry II
Possessions of the King of France
Dependencies of the King of France
0
100
Miles

22

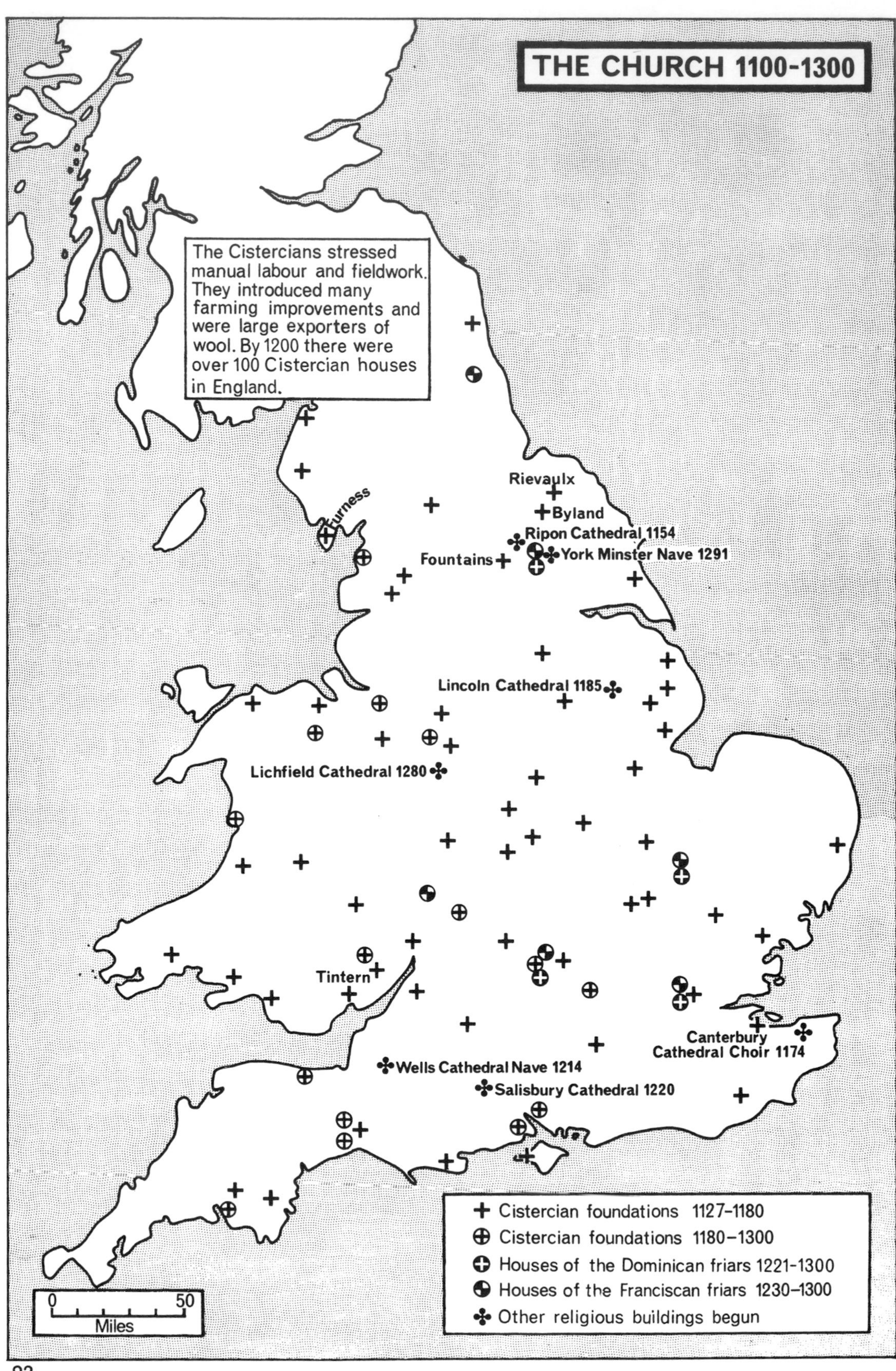
THE CHURCH 1100-1300
The Cistercians stressed manual labour and fieldwork. They introduced many farming improvements and were large exporters of wool. By 1200 there were over 100 Cistercian houses in England.
Rievaulx
Byland
Furness
Ripon Cathedral 1154
York Minster Nave 1291
Fountains
Lincoln Cathedral 1185
Lichfield Cathedral 1280
Tintern
Canterbury Cathedral Choir 1174
Wells Cathedral Nave 1214
Salisbury Cathedral 1220
Cistercian foundations 1127–1180
Cistercian foundations 1180–1300
Houses of the Dominican friars 1221-1300
Houses of the Franciscan friars 1230–1300
Other religious buildings begun
0
50
Miles

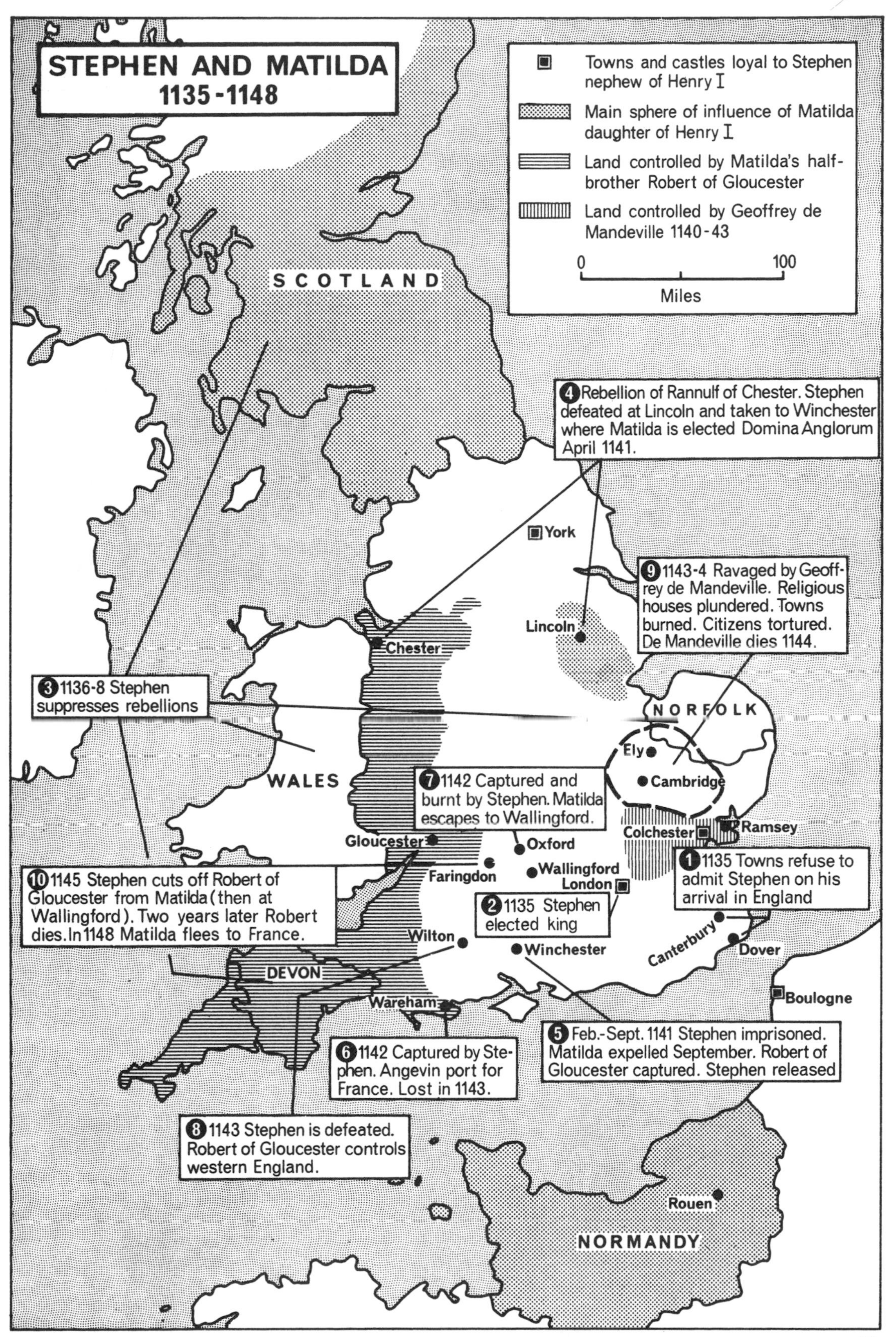
STEPHEN AND MATILDA
1135-1148
Towns and castles loyal to Stephen nephew of Henry I
Main sphere of influence of Matilda daughter of Henry I
Land controlled by Matilda's half-brother Robert of Gloucester
Land controlled by Geoffrey de Mandeville 1140-43
0
100
Miles
SCOTLAND
WALES
NORFOLK
DEVON
NORMANDY
York
Lincoln
Chester
Ely
Cambridge
Colchester
Ramsey
Gloucester
Oxford
Faringdon
Wallingford
London
Wilton
Winchester
Wareham
Canterbury
Dover
Boulogne
Rouen
1 1135 Towns refuse to admit Stephen on his arrival in England
2 1135 Stephen elected king
3 1136-8 Stephen suppresses rebellions
4 Rebellion of Rannulf of Chester. Stephen defeated at Lincoln and taken to Winchester where Matilda is elected Domina Anglorum April 1141.
5 Feb.-Sept. 1141 Stephen imprisoned. Matilda expelled September. Robert of Gloucester captured. Stephen released
6 1142 Captured by Stephen. Angevin port for France. Lost in 1143.
7 1142 Captured and burnt by Stephen. Matilda escapes to Wallingford.
8 1143 Stephen is defeated. Robert of Gloucester controls western England.
9 1143-4 Ravaged by Geoffrey de Mandeville. Religious houses plundered. Towns burned. Citizens tortured. De Mandeville dies 1144.
10 1145 Stephen cuts off Robert of Gloucester from Matilda (then at Wallingford). Two years later Robert dies. In 1148 Matilda flees to France.

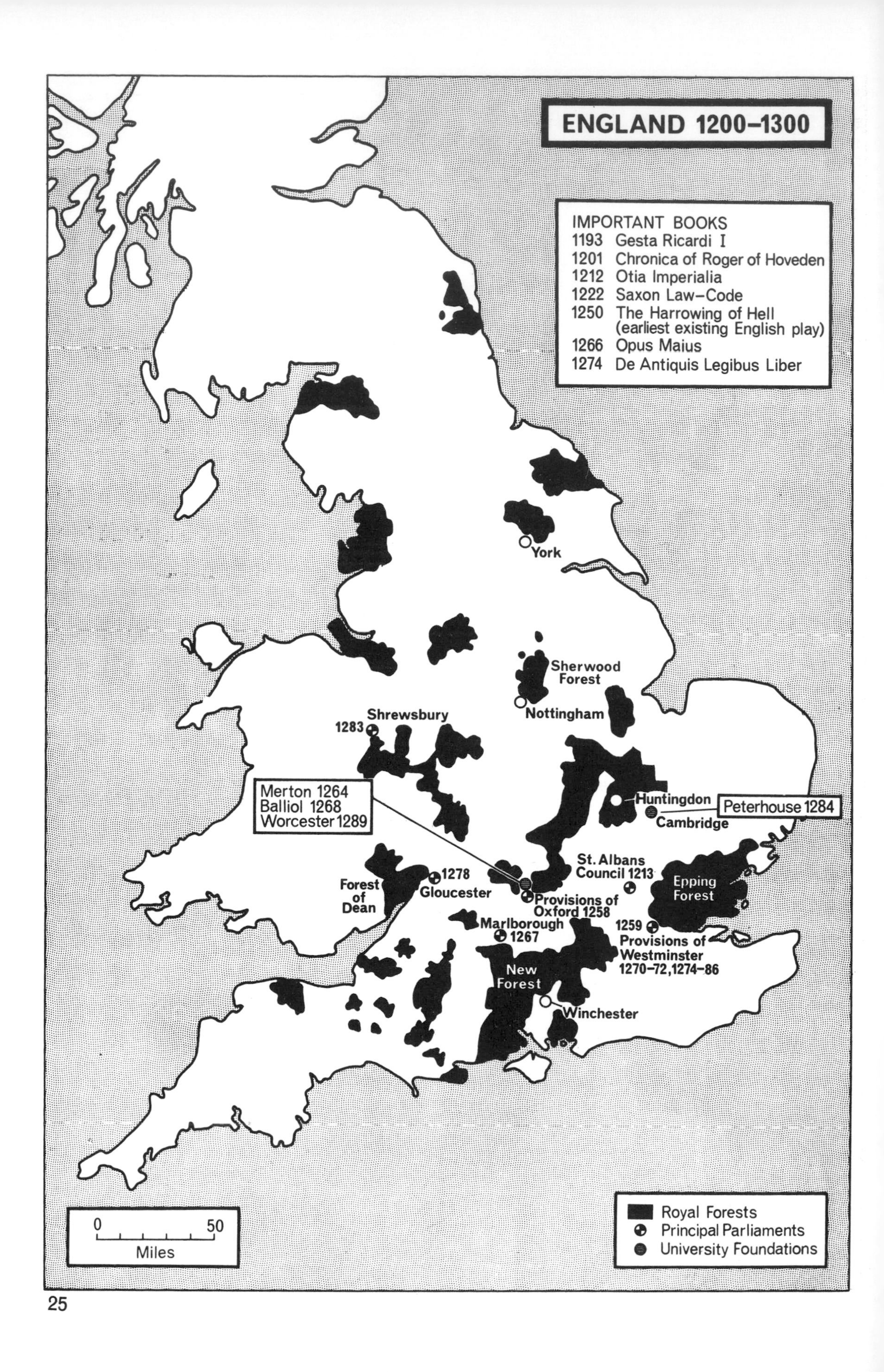
ENGLAND 1200–1300
IMPORTANT BOOKS
1193 Gesta Ricardi I
1201 Chronica of Roger of Hoveden
1212 Otia Imperialia
1222 Saxon Law-Code
1250 The Harrowing of Hell (earliest existing English play)
1266 Opus Maius
1274 De Antiquis Legibus Liber
York
Sherwood Forest
Nottingham
Shrewsbury
1283
Merton 1264
Balliol 1268
Worcester 1289
Huntingdon
Peterhouse 1284
Cambridge
St. Albans
Council 1213
Epping Forest
Forest of Dean
1278
Gloucester
Provisions of Oxford 1258
Marlborough
1267
1259
Provisions of Westminster
1270–72, 1274–86
New Forest
Winchester
0
50
Miles
Royal Forests
Principal Parliaments
University Foundations

THE ECONOMY 1200–1300
1245 Papal money-raiser expelled from England by king, clergy and barons
1274 Anglo-Flanders Commercial Treaty
1275 King to receive duty on wool
1280 German merchants in England form a Hansa
1290 Expulsion of the Jews from England
1299 Act to repress bad coinage passed
York
blues
Beverley
Lincoln scarlets
Lincoln
Nottingham
Stamfords
Norwich
Stamford
Leicester
Somersham
Coventry
Huntingdon
Ramsey
Northampton
Bury St.Edmunds
Worcester
Warwick
Cambridge
Ipswich
Bedford
Sudbury
russets
russets
Colchester
Gloucester
Oxford
russets
Chepstow
Wallingford
London
Bristol
Marlborough
Sandwich
Canterbury
Devizes
russets
Hythe
Dover
Romney
Rye
Wilton
Winchester
Winchelsea
Hastings
Salisbury
0
50
Miles
Cloth producing areas with names of cloth
Towns with weavers guilds by 1200
The Cinque Ports: special liberties granted 1278
The liberties of Chepstow, Ramsey and Somersham
Towns with Jewish settlements where Jewish loans were recorded 1190–1290

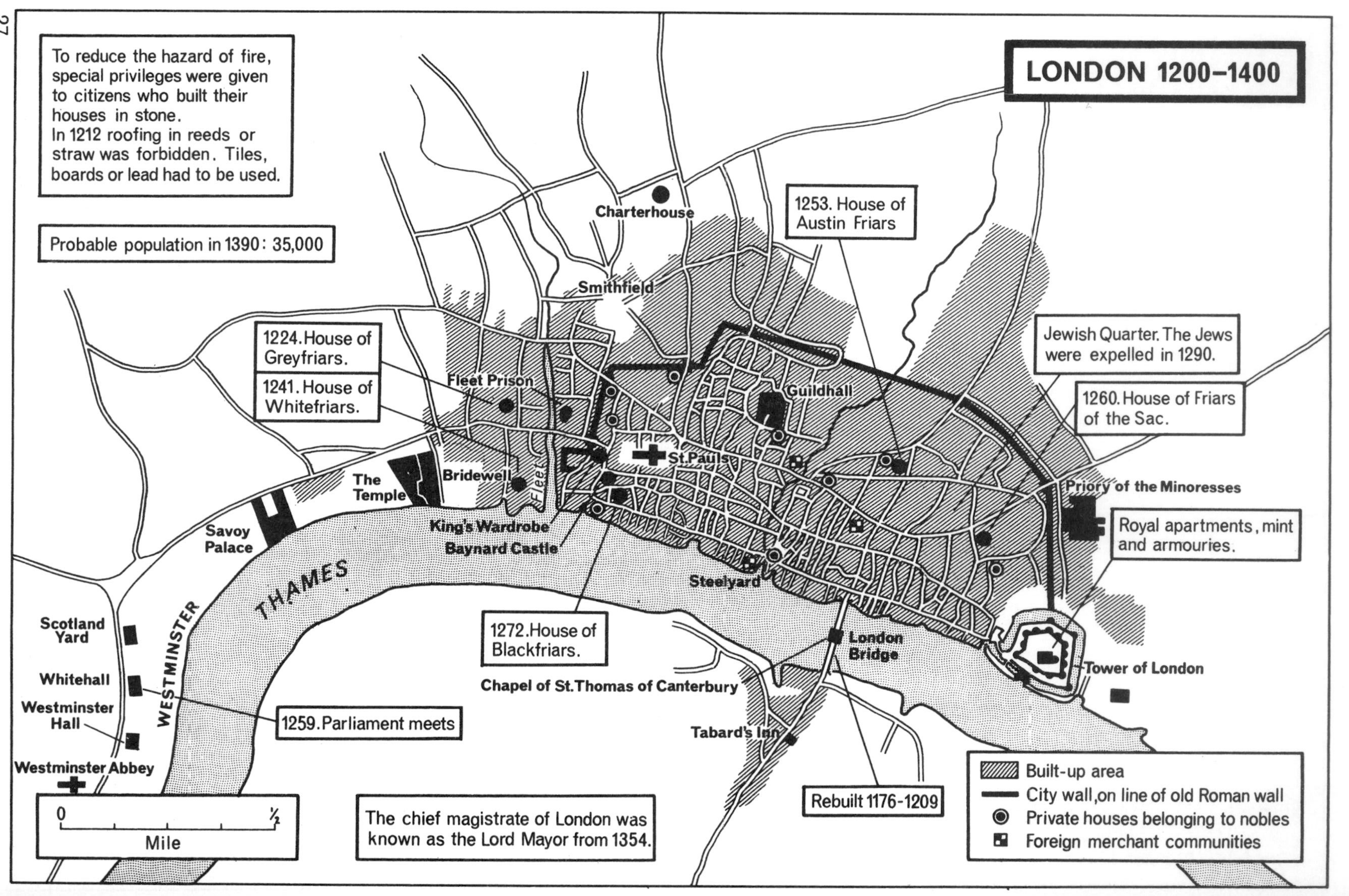
LONDON 1200–1400
To reduce the hazard of fire, special privileges were given to citizens who built their houses in stone.
In 1212 roofing in reeds or straw was forbidden. Tiles, boards or lead had to be used.
Probable population in 1390: 35,000
Charterhouse
1253. House of Austin Friars
Smithfield
1224. House of Greyfriars.
1241. House of Whitefriars.
Fleet Prison
Guildhall
Jewish Quarter. The Jews were expelled in 1290.
1260. House of Friars of the Sac.
St. Pauls
Bridewell
The Temple
Fleet
Priory of the Minoresses
Royal apartments, mint and armouries.
Savoy Palace
King's Wardrobe
Baynard Castle
THAMES
Steelyard
WESTMINSTER
Scotland Yard
Whitehall
Westminster Hall
Westminster Abbey
1272. House of Blackfriars.
London Bridge
Tower of London
Chapel of St. Thomas of Canterbury
1259. Parliament meets
Tabard's Inn
0
½
Mile
The chief magistrate of London was known as the Lord Mayor from 1354.
Rebuilt 1176-1209
Built-up area
City wall, on line of old Roman wall
Private houses belonging to nobles
Foreign merchant communities

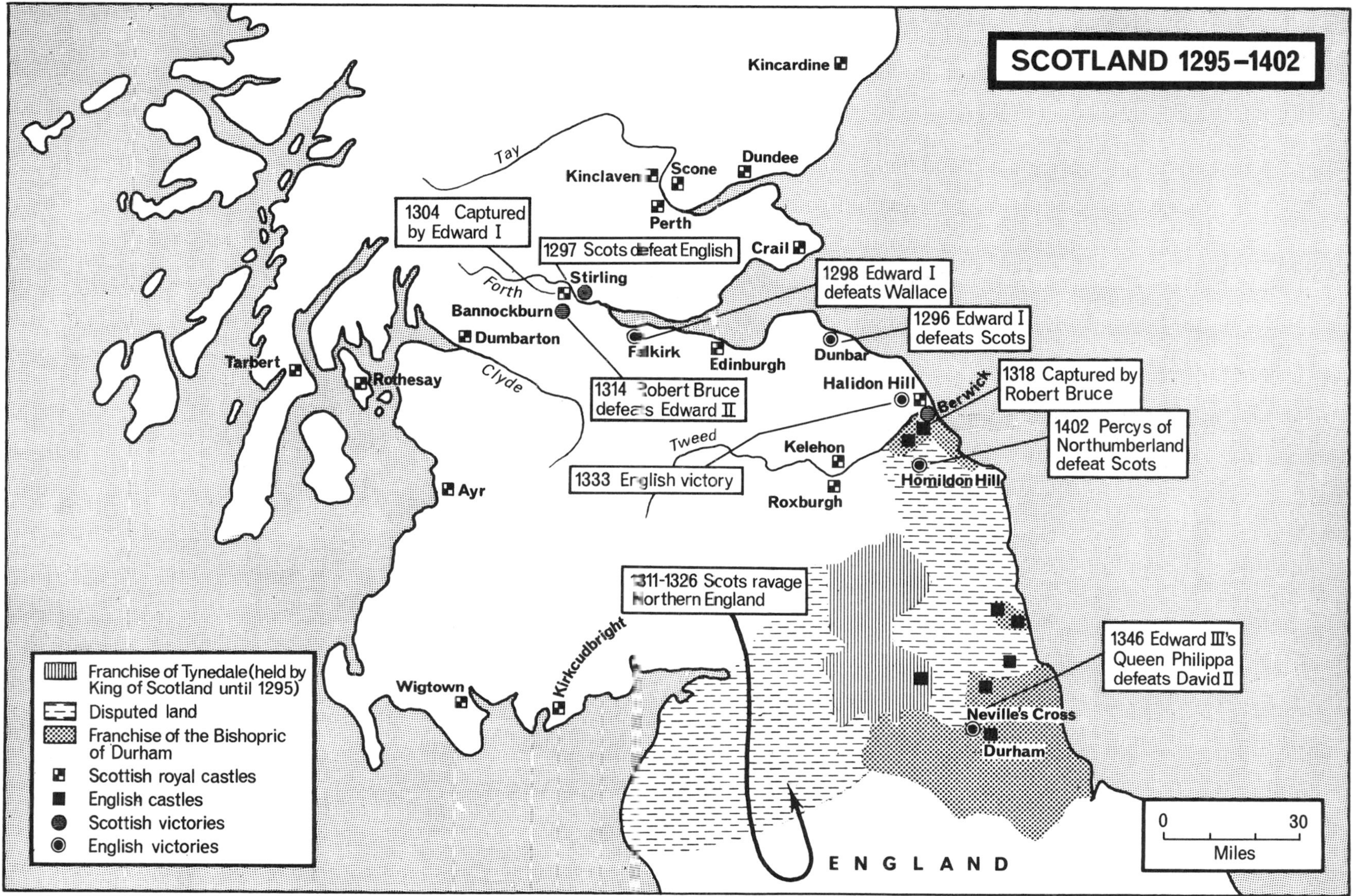
SCOTLAND 1295–1402
Kincardine
Tay
Kinclaven
Scone
Dundee
Perth
Crail
1304 Captured by Edward I
1297 Scots defeat English
Stirling
Forth
Bannockburn
Dumbarton
1298 Edward I defeats Wallace
1296 Edward I defeats Scots
Falkirk
Edinburgh
Dunbar
Tarbert
Rothesay
Clyde
1314 Robert Bruce defeats Edward II
Halidon Hill
Berwick
1318 Captured by Robert Bruce
1402 Percys of Northumberland defeat Scots
Tweed
Kelehon
1333 English victory
Homildon Hill
Roxburgh
Ayr
1311-1326 Scots ravage Northern England
Kirkcudbright
Wigtown
1346 Edward III's Queen Philippa defeats David II
Neville's Cross
Durham
ENGLAND
0
30
Miles
Franchise of Tynedale (held by King of Scotland until 1295)
Disputed land
Franchise of the Bishopric of Durham
Scottish royal castles
English castles
Scottish victories
English victories

c

THE HUNDRED YEARS' WAR
1259-1368
0
100
Miles
Calais
Etaples
Crécy
Abbeville
HOLY
Barfleur
Rouen
Caen
NORMANDY
Paris
ROMAN
Bretigny
F
R
A
N
C
E
ANJOU
Tours
EMPIRE
Bourges
Poitiers
POITOU
AQUITAINE
Bordeaux
GUYENNE
DAUPHINÉ
QUERCY
ROUERGUE
GASCONY
Bayonne
Toulouse
Vitoria
Narbonne
Pass of Roncesvalles
Pamplona
To Burgos
NAVARRE
ARAGON
Possessions of Henry III, 1259
Possessions of the King of France, 1259
English gains 1275
English gains at the Treaty of Bretigny, 1368
Edward III's campaign 1346-1349
the three campaigns of Edward the Black Prince:
to Narbonne 1355
to Poitiers 1356
to Burgos 1367

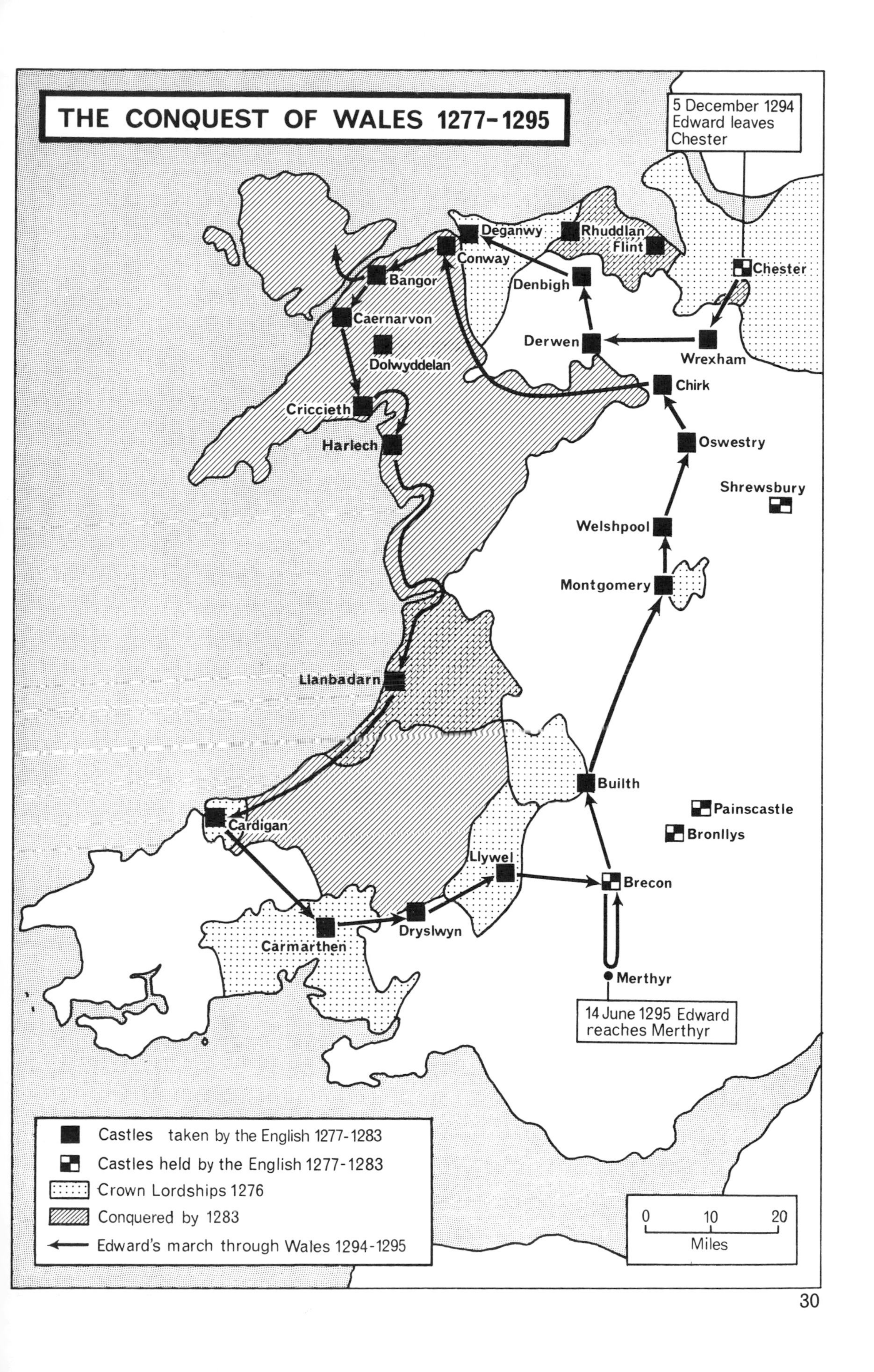
THE CONQUEST OF WALES 1277-1295
5 December 1294
Edward leaves
Chester
Deganwy
Rhuddlan
Flint
Conway
Chester
Bangor
Denbigh
Caernarvon
Derwen
Wrexham
Dolwyddelan
Chirk
Criccieth
Harlech
Oswestry
Shrewsbury
Welshpool
Montgomery
Llanbadarn
Builth
Painscastle
Cardigan
Bronllys
Llywel
Brecon
Dryslwyn
Carmarthen
Merthyr
14 June 1295 Edward
reaches Merthyr
Castles taken by the English 1277-1283
Castles held by the English 1277-1283
Crown Lordships 1276
Conquered by 1283
Edward's march through Wales 1294-1295
0
10
20
Miles

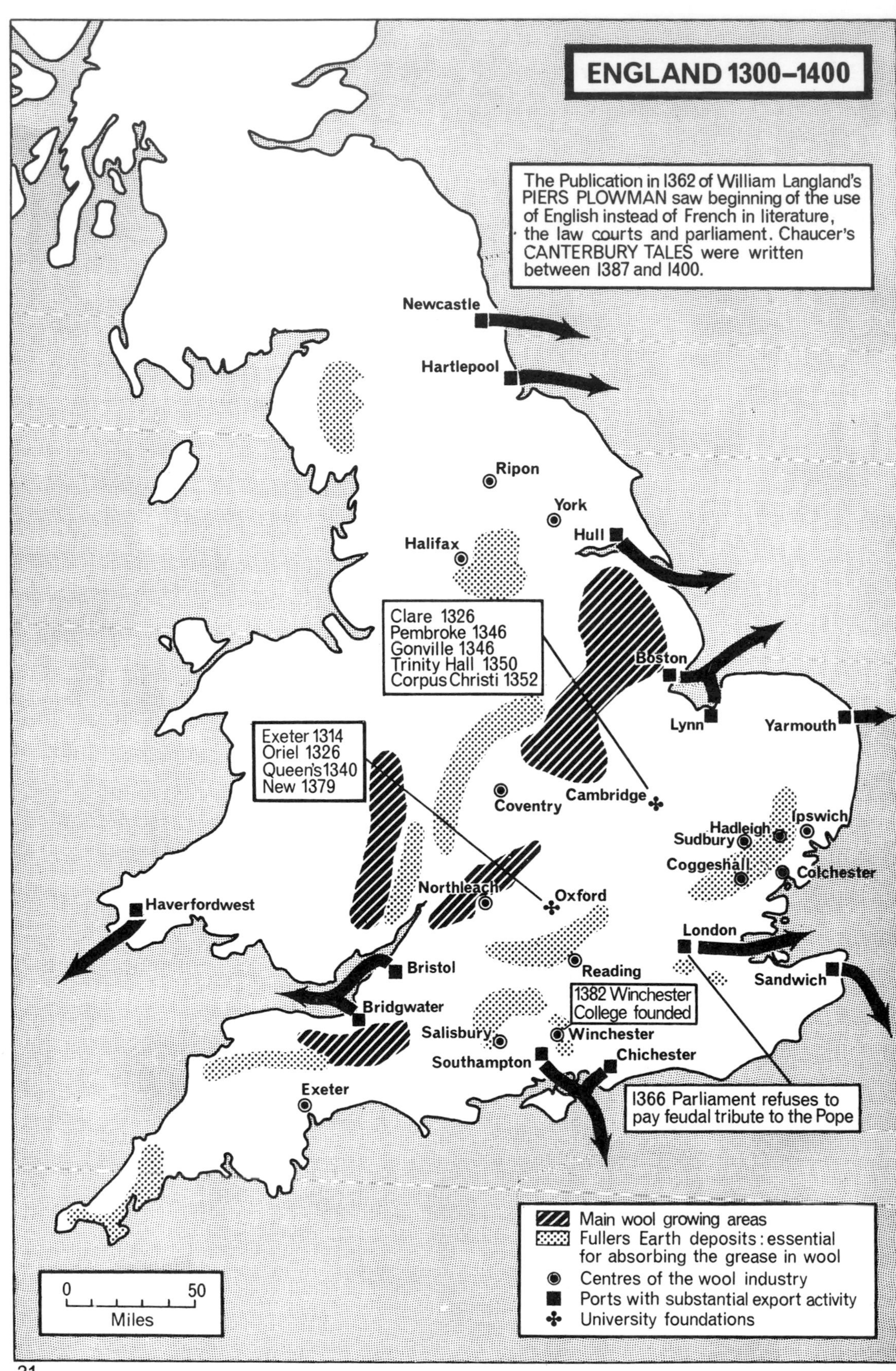
ENGLAND 1300–1400
The Publication in 1362 of William Langland's PIERS PLOWMAN saw beginning of the use of English instead of French in literature, the law courts and parliament. Chaucer's CANTERBURY TALES were written between 1387 and 1400.
Newcastle
Hartlepool
Ripon
York
Hull
Halifax
Clare 1326
Pembroke 1346
Gonville 1346
Trinity Hall 1350
Corpus Christi 1352
Boston
Lynn
Yarmouth
Exeter 1314
Oriel 1326
Queen's 1340
New 1379
Coventry
Cambridge
Hadleigh
Ipswich
Sudbury
Coggeshall
Colchester
Northleach
Oxford
Haverfordwest
London
Bristol
Reading
Sandwich
Bridgwater
1382 Winchester College founded
Salisbury
Winchester
Chichester
Southampton
Exeter
1366 Parliament refuses to pay feudal tribute to the Pope
Main wool growing areas
Fullers Earth deposits: essential for absorbing the grease in wool
Centres of the wool industry
Ports with substantial export activity
University foundations
0
50
Miles

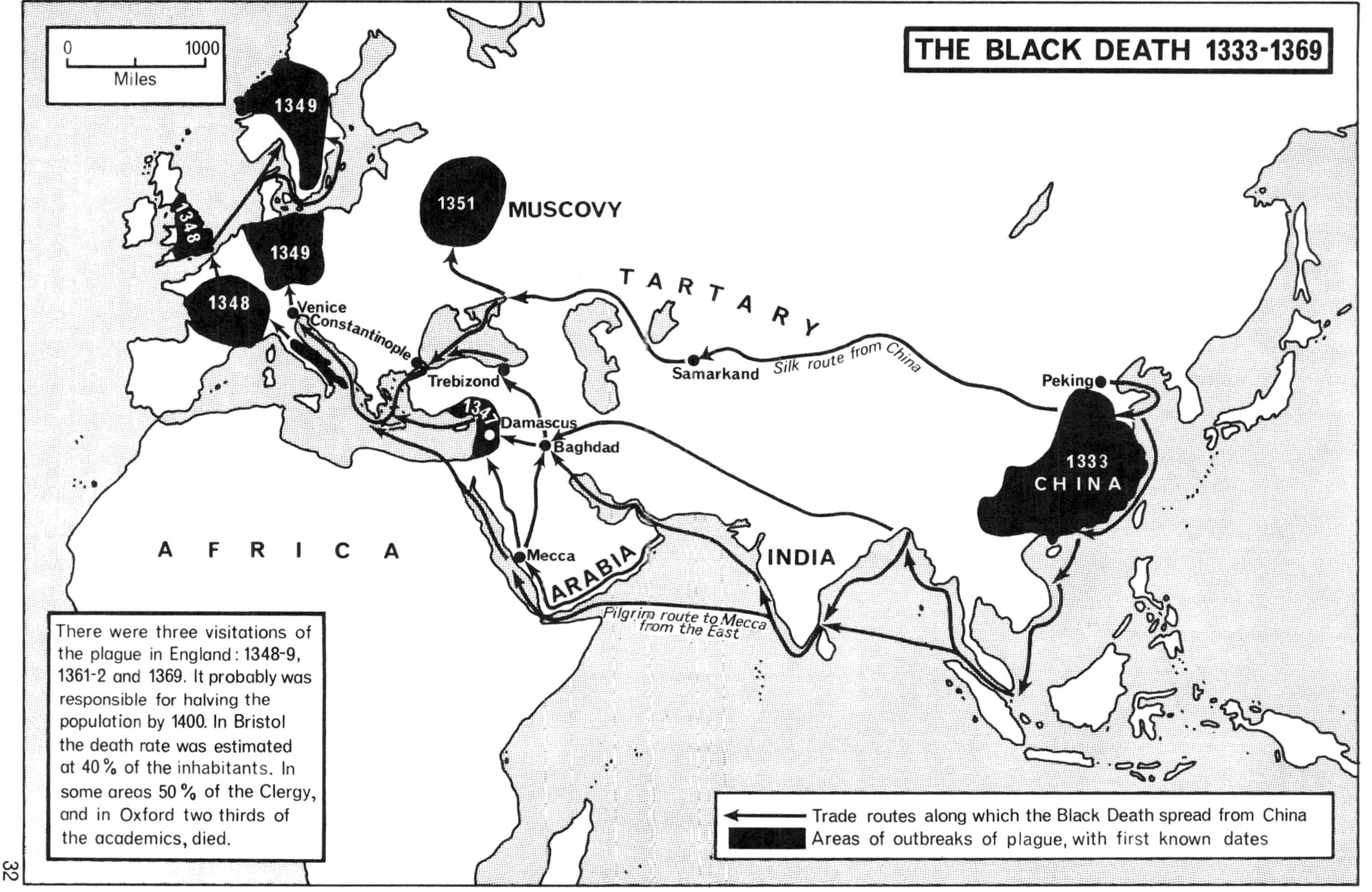
THE BLACK DEATH 1333-1369
0
1000
Miles
1349
1348
1349
1348
1351
MUSCOVY
TARTARY
Venice
Constantinople
Trebizond
Samarkand
Silk route from China
Peking
1333
CHINA
1347
Damascus
Baghdad
Mecca
ARABIA
INDIA
AFRICA
Pilgrim route to Mecca from the East
There were three visitations of the plague in England: 1348-9, 1361-2 and 1369. It probably was responsible for halving the population by 1400. In Bristol the death rate was estimated at 40 % of the inhabitants. In some areas 50 % of the Clergy, and in Oxford two thirds of the academics, died.
Trade routes along which the Black Death spread from China
Areas of outbreaks of plague, with first known dates

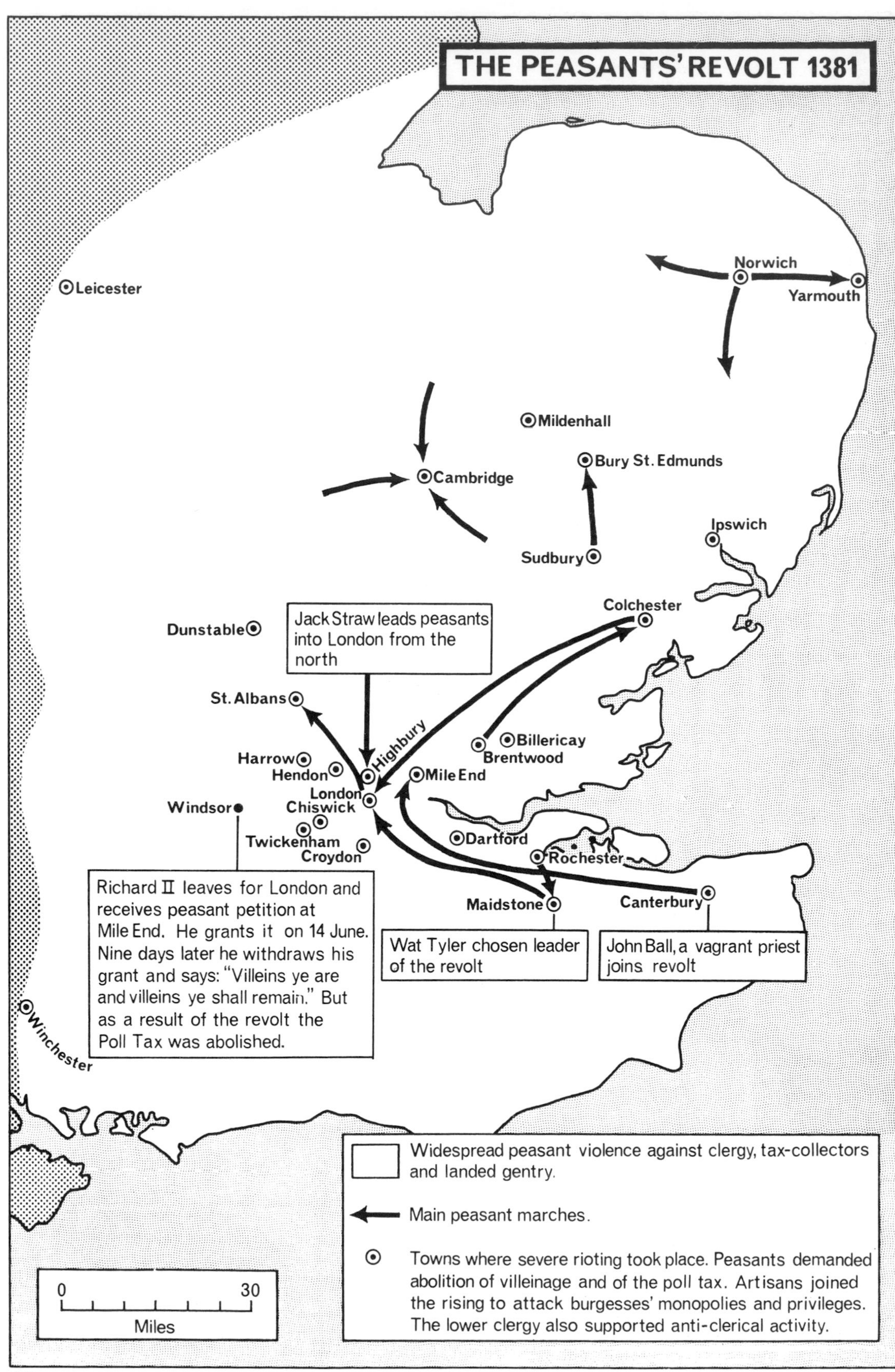
THE PEASANTS' REVOLT 1381
Norwich
Yarmouth
Leicester
Mildenhall
Bury St. Edmunds
Cambridge
Ipswich
Sudbury
Colchester
Dunstable
Jack Straw leads peasants into London from the north
St. Albans
Highbury
Billericay
Brentwood
Harrow
Hendon
Mile End
London
Windsor
Chiswick
Twickenham
Croydon
Dartford
Rochester
Maidstone
Canterbury
Richard II leaves for London and receives peasant petition at Mile End. He grants it on 14 June. Nine days later he withdraws his grant and says: "Villeins ye are and villeins ye shall remain." But as a result of the revolt the Poll Tax was abolished.
Wat Tyler chosen leader of the revolt
John Ball, a vagrant priest joins revolt
Winchester
Widespread peasant violence against clergy, tax-collectors and landed gentry.
Main peasant marches.
Towns where severe rioting took place. Peasants demanded abolition of villeinage and of the poll tax. Artisans joined the rising to attack burgesses' monopolies and privileges. The lower clergy also supported anti-clerical activity.
0
30
Miles

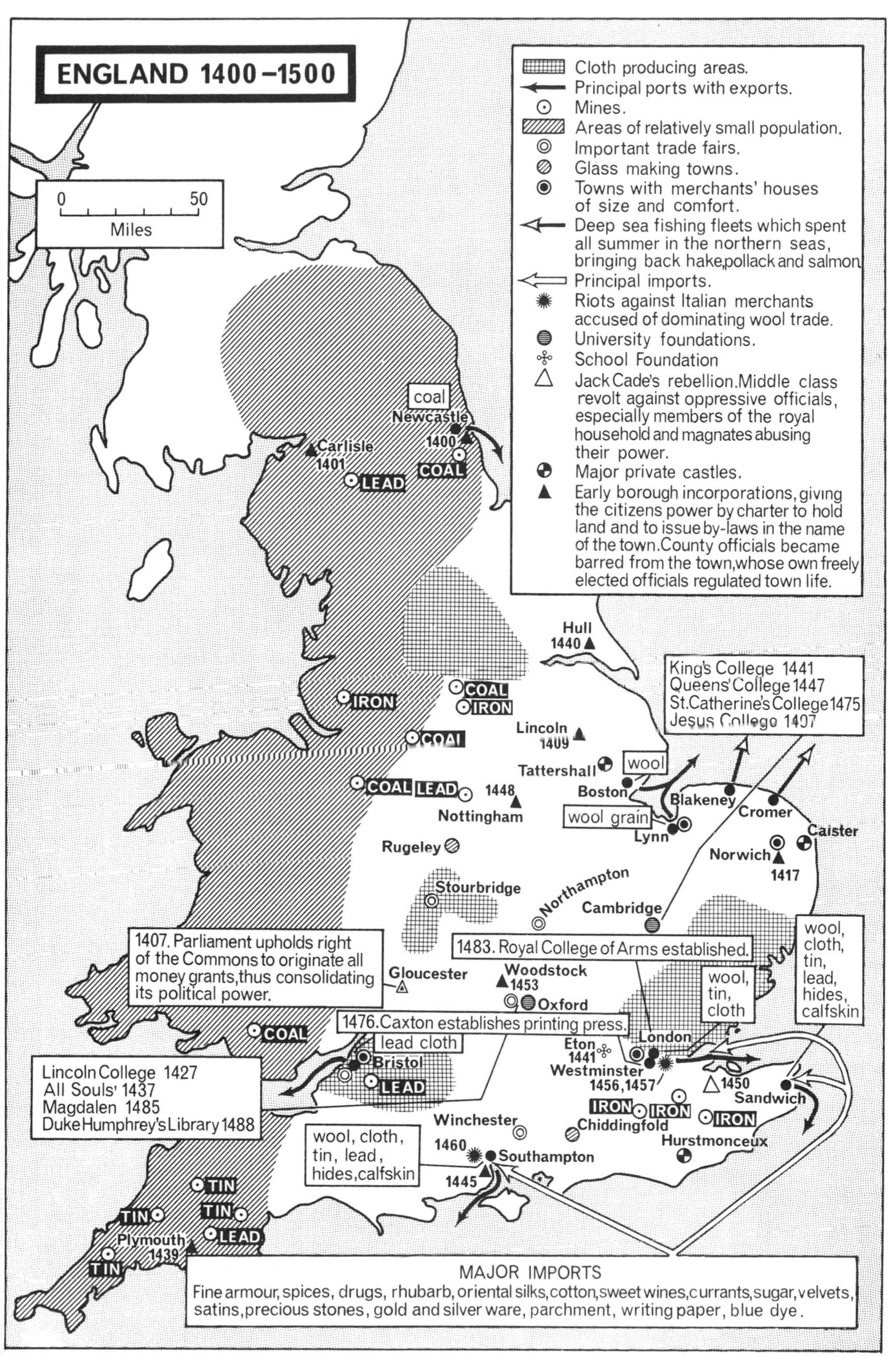
ENGLAND 1400–1500
Cloth producing areas.
Principal ports with exports.
Mines.
Areas of relatively small population.
Important trade fairs.
Glass making towns.
Towns with merchants' houses of size and comfort.
Deep sea fishing fleets which spent all summer in the northern seas, bringing back hake,pollack and salmon
Principal imports.
Riots against Italian merchants accused of dominating wool trade.
University foundations.
School Foundation
Jack Cade's rebellion.Middle class revolt against oppressive officials, especially members of the royal household and magnates abusing their power.
Major private castles.
Early borough incorporations, giving the citizens power by charter to hold land and to issue by-laws in the name of the town.County officials became barred from the town,whose own freely elected officials regulated town life.
0
50
Miles
coal
Newcastle
1400
COAL
Carlisle
1401
LEAD
Hull
1440
King's College 1441
Queens' College 1447
St.Catherine's College 1475
Jesus College 1497
COAL
IRON
IRON
Lincoln
1409
COAL
Tattershall
wool
Boston
COAL
LEAD
1448
Blakeney
Cromer
Nottingham
wool grain
Lynn
Caister
Rugeley
Norwich
1417
Stourbridge
Northampton
Cambridge
1407. Parliament upholds right of the Commons to originate all money grants,thus consolidating its political power.
1483. Royal College of Arms established.
wool, cloth, tin, lead, hides, calfskin
Gloucester
Woodstock
1453
wool, tin, cloth
Oxford
1476.Caxton establishes printing press.
COAL
lead cloth
London
Bristol
Eton
1441
Lincoln College 1427
All Souls' 1437
Magdalen 1485
Duke Humphrey's Library 1488
LEAD
Westminster
1456,1457
1450
Sandwich
IRON
IRON
IRON
Winchester
Chiddingfold
wool, cloth, tin, lead, hides,calfskin
1460
Southampton
Hurstmonceux
1445
TIN
TIN
TIN
LEAD
Plymouth
1439
TIN
MAJOR IMPORTS
Fine armour, spices, drugs, rhubarb, oriental silks, cotton, sweet wines, currants, sugar, velvets, satins, precious stones, gold and silver ware, parchment, writing paper, blue dye.

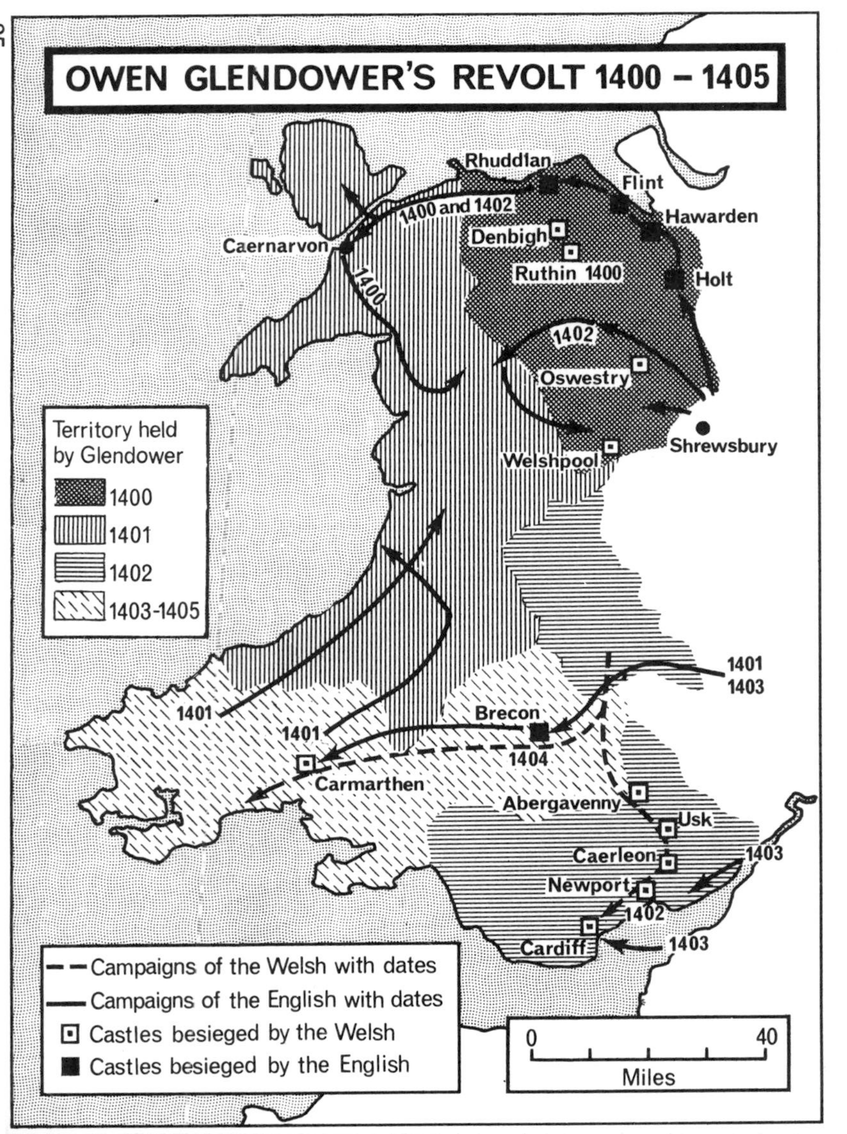
35
OWEN GLENDOWER'S REVOLT 1400 – 1405
Rhuddlan
Flint
Hawarden
1400 and 1402
Caernarvon
Denbigh
Ruthin 1400
Holt
1400
1402
Oswestry
Shrewsbury
Welshpool
Territory held by Glendower
1400
1401
1402
1403-1405
1401
1403
1401
1401
Brecon
1404
Carmarthen
Abergavenny
Usk
Caerleon
1403
Newport
1402
Cardiff
1403
Campaigns of the Welsh with dates
Campaigns of the English with dates
Castles besieged by the Welsh
Castles besieged by the English
0
40
Miles

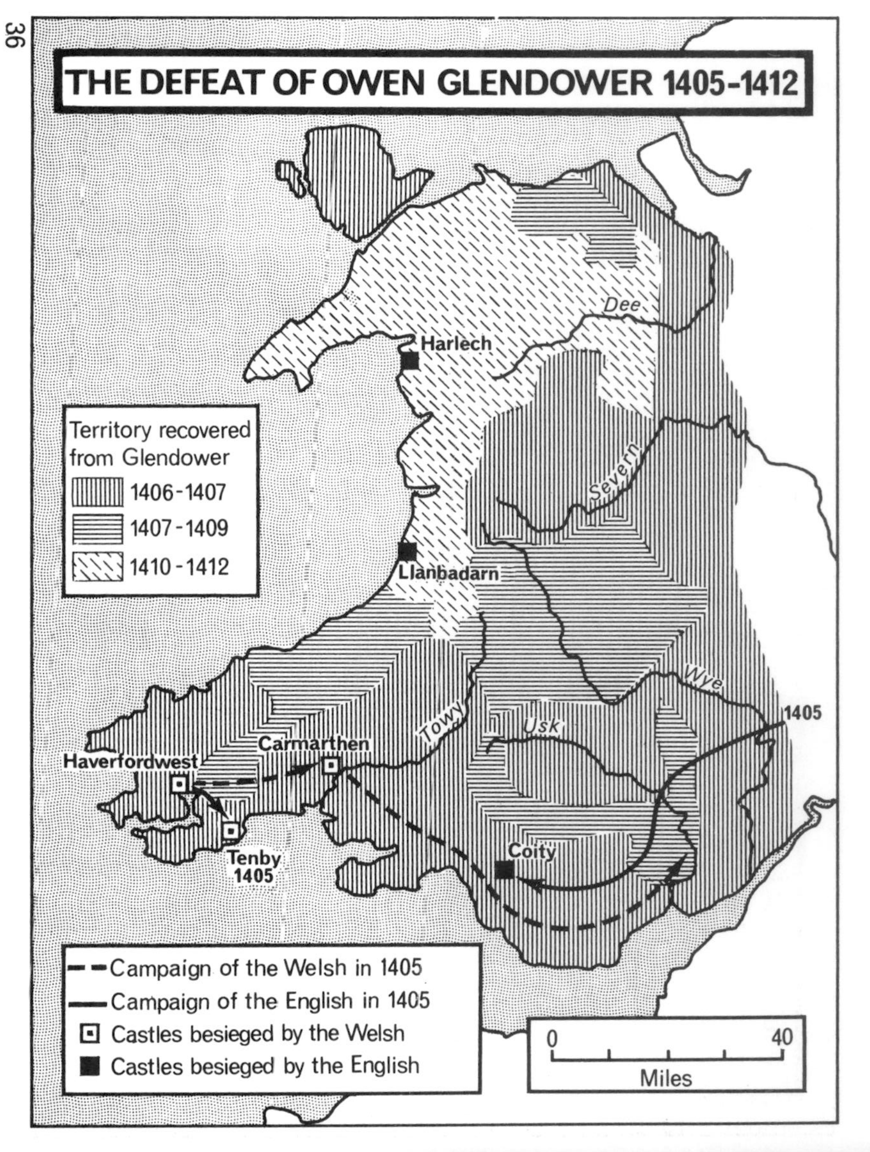
36
THE DEFEAT OF OWEN GLENDOWER 1405-1412
Dee
Harlech
Territory recovered from Glendower
1406-1407
1407-1409
1410-1412
Severn
Llanbadarn
Wye
Towy
Usk
1405
Haverfordwest
Carmarthen
Tenby 1405
Coity
Campaign of the Welsh in 1405
Campaign of the English in 1405
Castles besieged by the Welsh
Castles besieged by the English
0
40
Miles

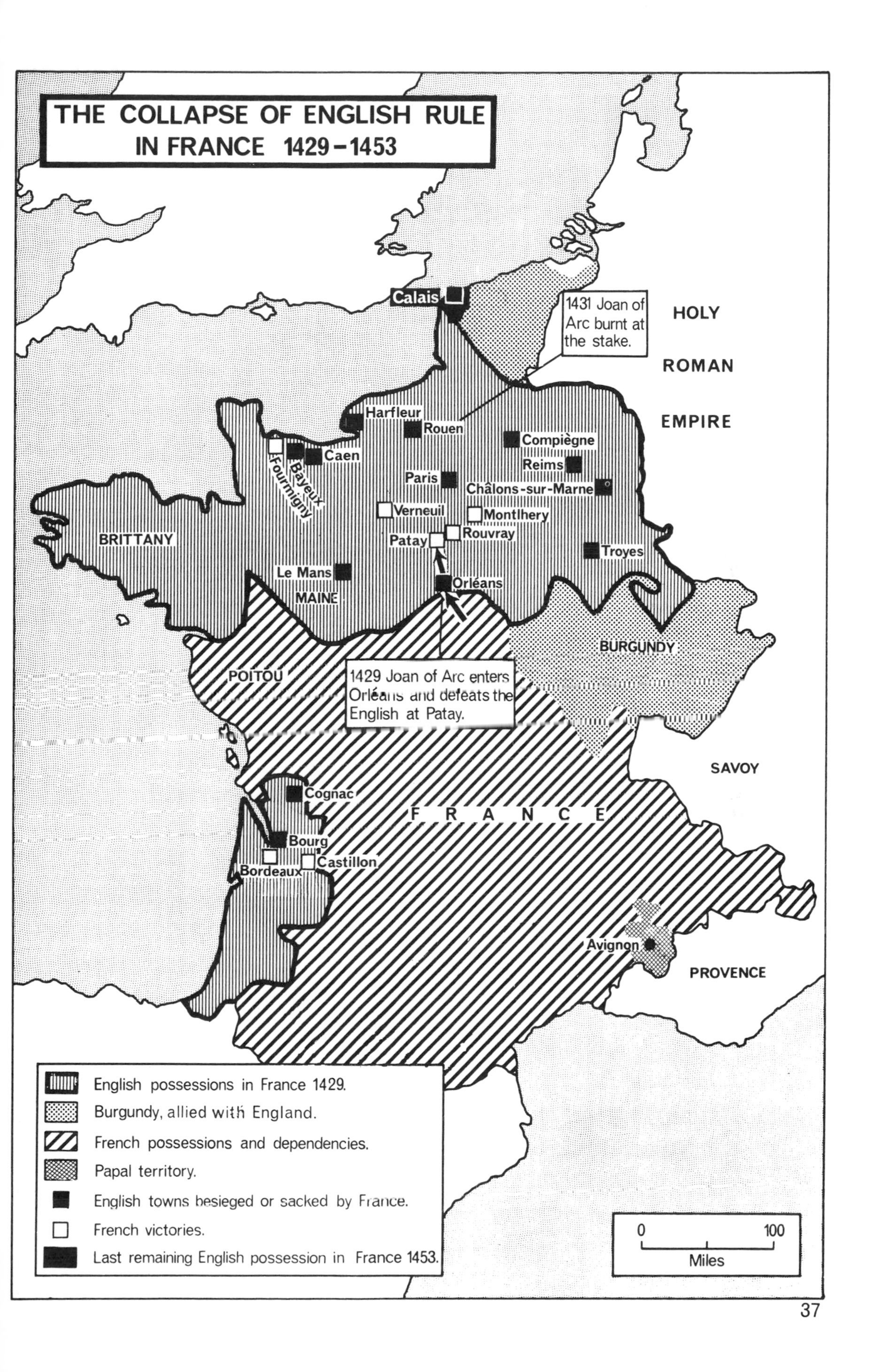
THE COLLAPSE OF ENGLISH RULE IN FRANCE 1429–1453
Calais
1431 Joan of Arc burnt at the stake.
HOLY
ROMAN
EMPIRE
Harfleur
Rouen
Compiègne
Reims
Caen
Bayeux
Fourmigny
Paris
Châlons-sur-Marne
Verneuil
Montlhery
Patay
Rouvray
Troyes
BRITTANY
Le Mans
Orléans
MAINE
BURGUNDY
POITOU
1429 Joan of Arc enters Orléans and defeats the English at Patay.
SAVOY
Cognac
F R A N C E
Bourg
Castillon
Bordeaux
Avignon
PROVENCE
English possessions in France 1429.
Burgundy, allied with England.
French possessions and dependencies.
Papal territory.
English towns besieged or sacked by France.
French victories.
Last remaining English possession in France 1453.
0
100
Miles

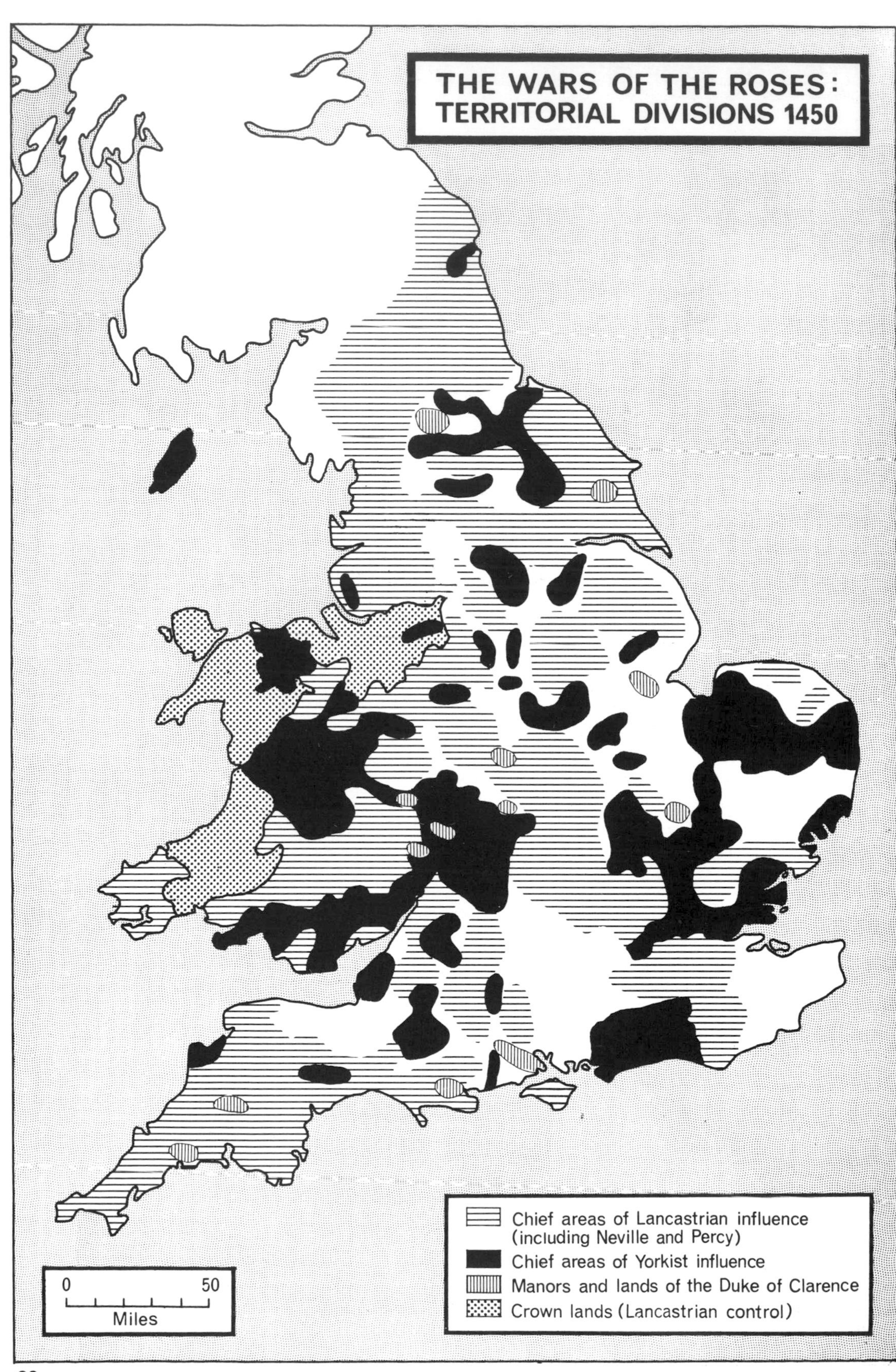
THE WARS OF THE ROSES:
TERRITORIAL DIVISIONS 1450
Chief areas of Lancastrian influence (including Neville and Percy)
Chief areas of Yorkist influence
Manors and lands of the Duke of Clarence
Crown lands (Lancastrian control)
0
50
Miles

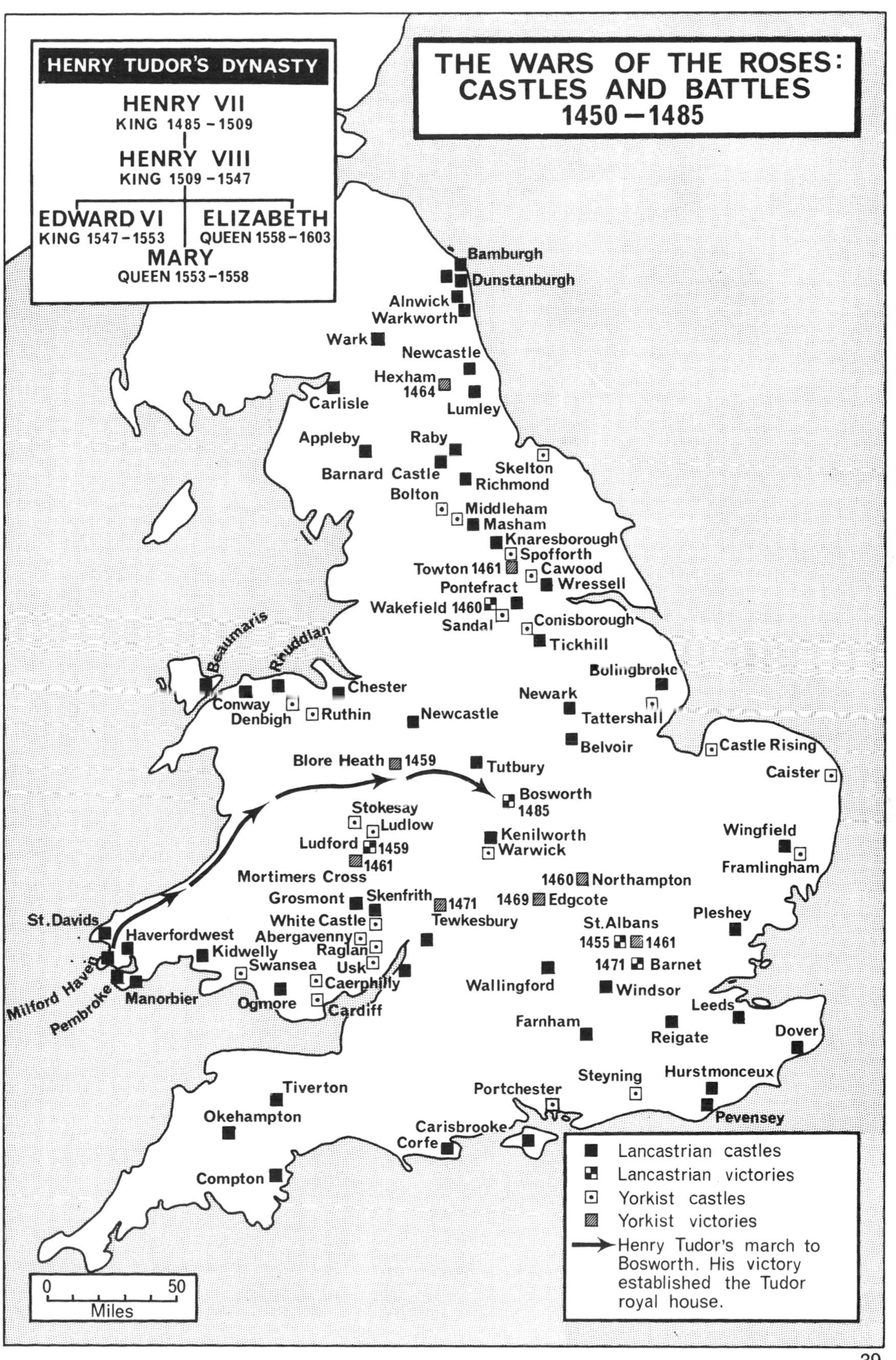
THE WARS OF THE ROSES: CASTLES AND BATTLES 1450—1485
HENRY TUDOR'S DYNASTY
HENRY VII
KING 1485 - 1509
HENRY VIII
KING 1509 - 1547
EDWARD VI
KING 1547 - 1553
ELIZABETH
QUEEN 1558 - 1603
MARY
QUEEN 1553 - 1558
Bamburgh
Dunstanburgh
Alnwick
Warkworth
Wark
Newcastle
Hexham 1464
Carlisle
Lumley
Appleby
Raby
Barnard Castle
Skelton
Richmond
Bolton
Middleham
Masham
Knaresborough
Spofforth
Towton 1461
Cawood
Pontefract
Wressell
Wakefield 1460
Sandal
Conisborough
Tickhill
Beaumaris
Rhuddlan
Chester
Conway
Denbigh
Ruthin
Newcastle
Newark
Tattershall
Belvoir
Castle Rising
Caister
Blore Heath 1459
Tutbury
Bosworth 1485
Stokesay
Ludlow
Ludford 1459
1461
Mortimers Cross
Kenilworth
Warwick
Wingfield
Framlingham
1460 Northampton
1469 Edgcote
Grosmont
Skenfrith
1471
Tewkesbury
White Castle
Abergavenny
Raglan
Usk
Caerphilly
Cardiff
Swansea
Kidwelly
Ogmore
St. Davids
Haverfordwest
Milford Haven
Pembroke
Manorbier
St. Albans
1455 1461
1471 Barnet
Pleshey
Wallingford
Windsor
Leeds
Farnham
Reigate
Dover
Steyning
Hurstmonceux
Portchester
Pevensey
Tiverton
Okehampton
Carisbrooke
Corfe
Compton
0 50
Miles
Lancastrian castles
Lancastrian victories
Yorkist castles
Yorkist victories
Henry Tudor's march to Bosworth. His victory established the Tudor royal house.

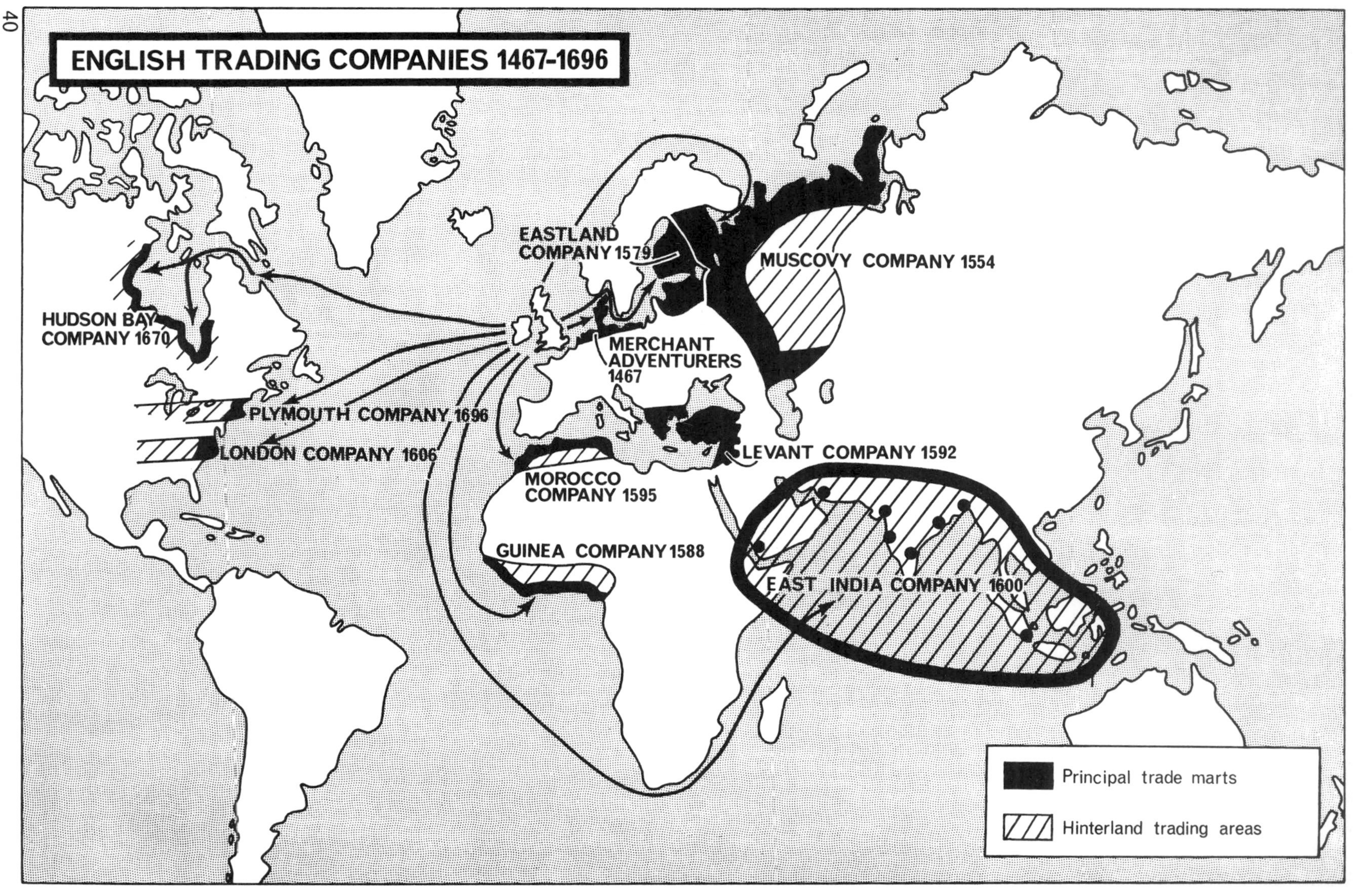
ENGLISH TRADING COMPANIES 1467-1696
HUDSON BAY COMPANY 1670
EASTLAND COMPANY 1579
MUSCOVY COMPANY 1554
MERCHANT ADVENTURERS 1467
PLYMOUTH COMPANY 1696
LONDON COMPANY 1606
LEVANT COMPANY 1592
MOROCCO COMPANY 1595
GUINEA COMPANY 1588
EAST INDIA COMPANY 1600
Principal trade marts
Hinterland trading areas

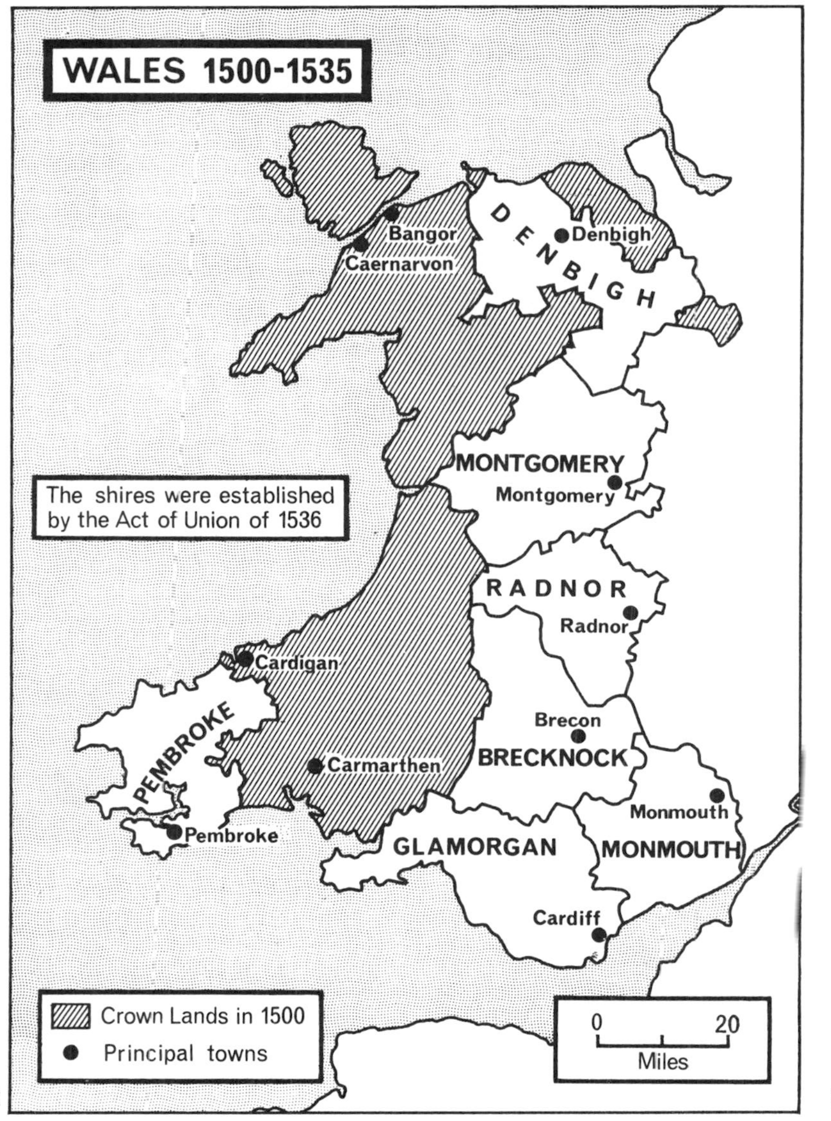

WALES 1500-1535
The shires were established by the Act of Union of 1536
DENBIGH
Denbigh
Bangor
Caernarvon
MONTGOMERY
Montgomery
RADNOR
Radnor
Cardigan
PEMBROKE
Pembroke
Carmarthen
Brecon
BRECKNOCK
Monmouth
MONMOUTH
GLAMORGAN
Cardiff
Crown Lands in 1500
Principal towns
0
20
Miles

41

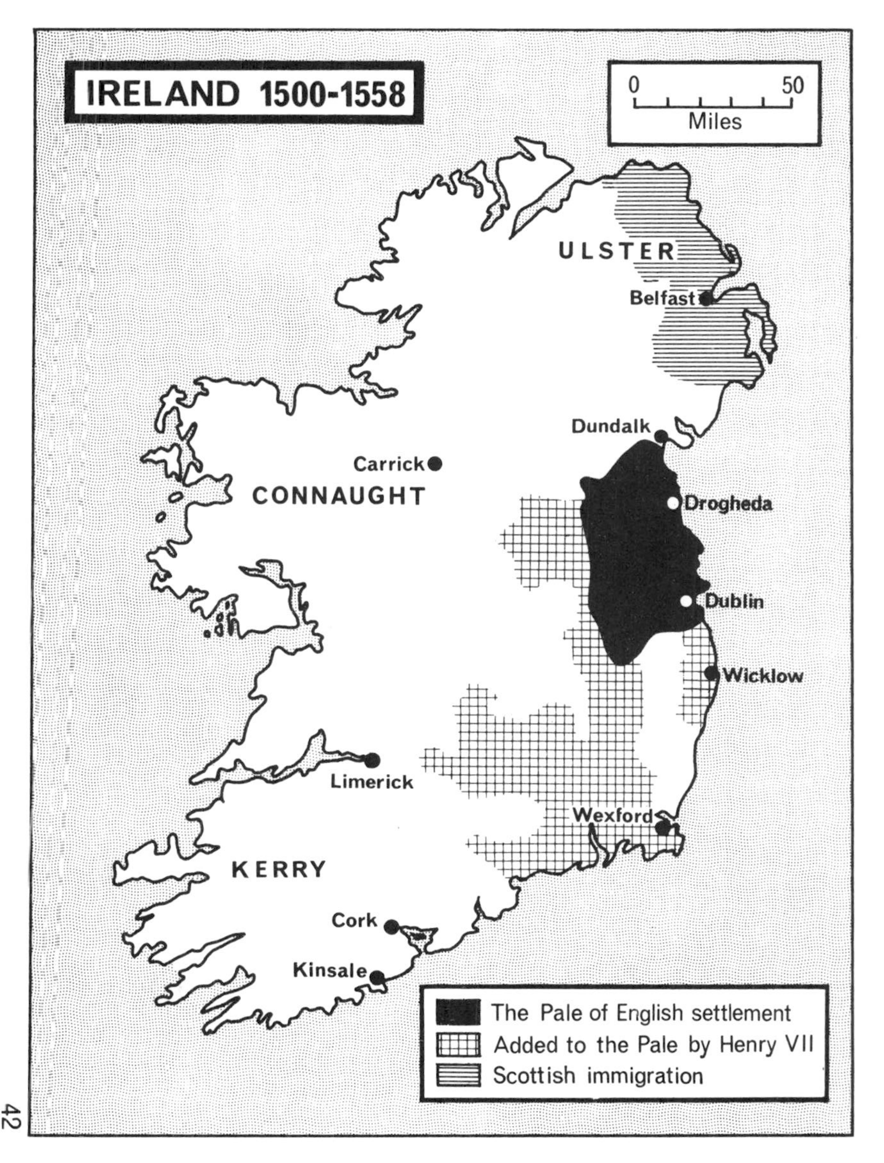

IRELAND 1500-1558
0
50
Miles
ULSTER
Belfast
Dundalk
Carrick
CONNAUGHT
Drogheda
Dublin
Wicklow
Limerick
Wexford
KERRY
Cork
Kinsale
The Pale of English settlement
Added to the Pale by Henry VII
Scottish immigration

42

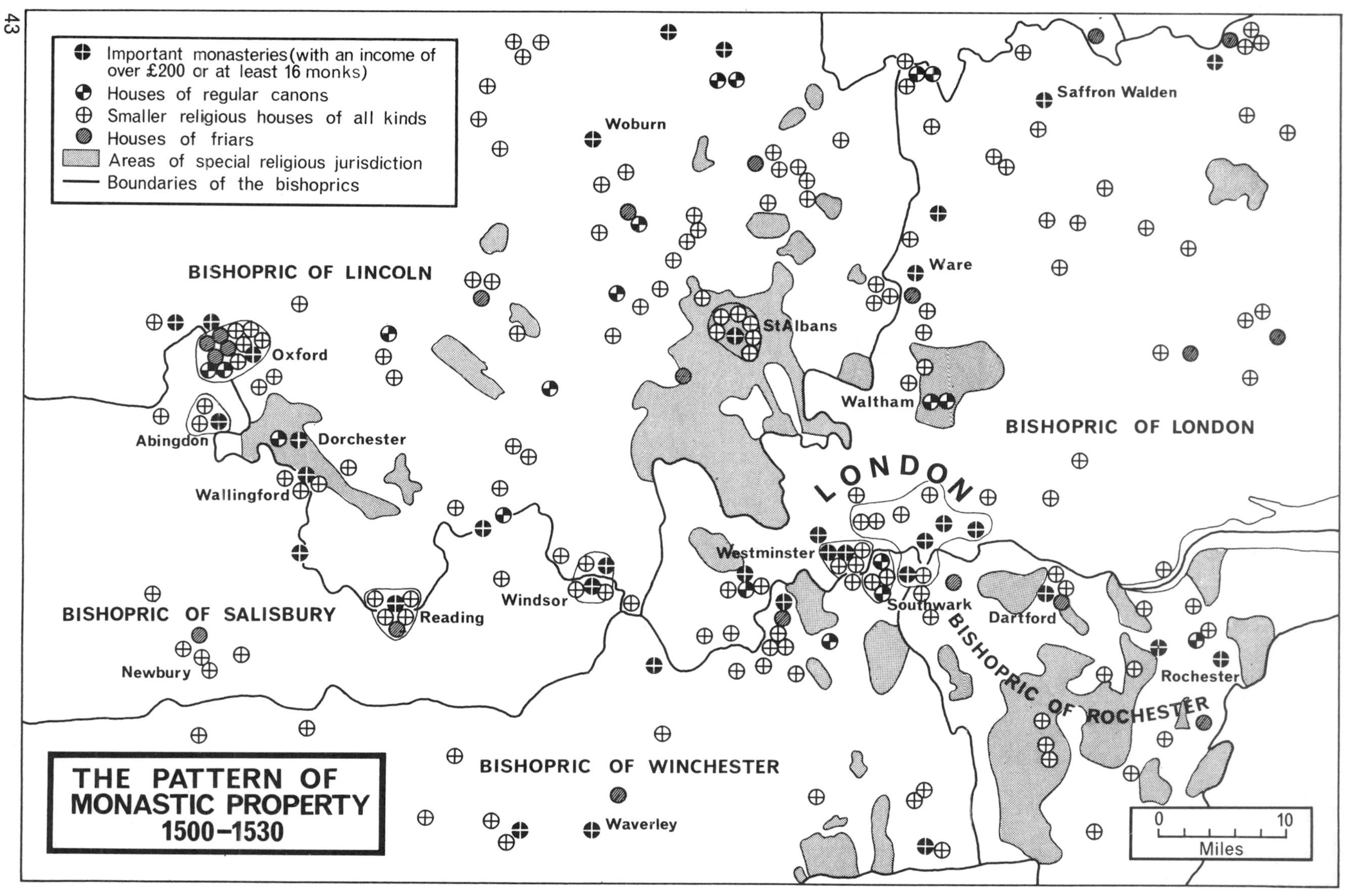

THE PATTERN OF MONASTIC PROPERTY 1500–1530

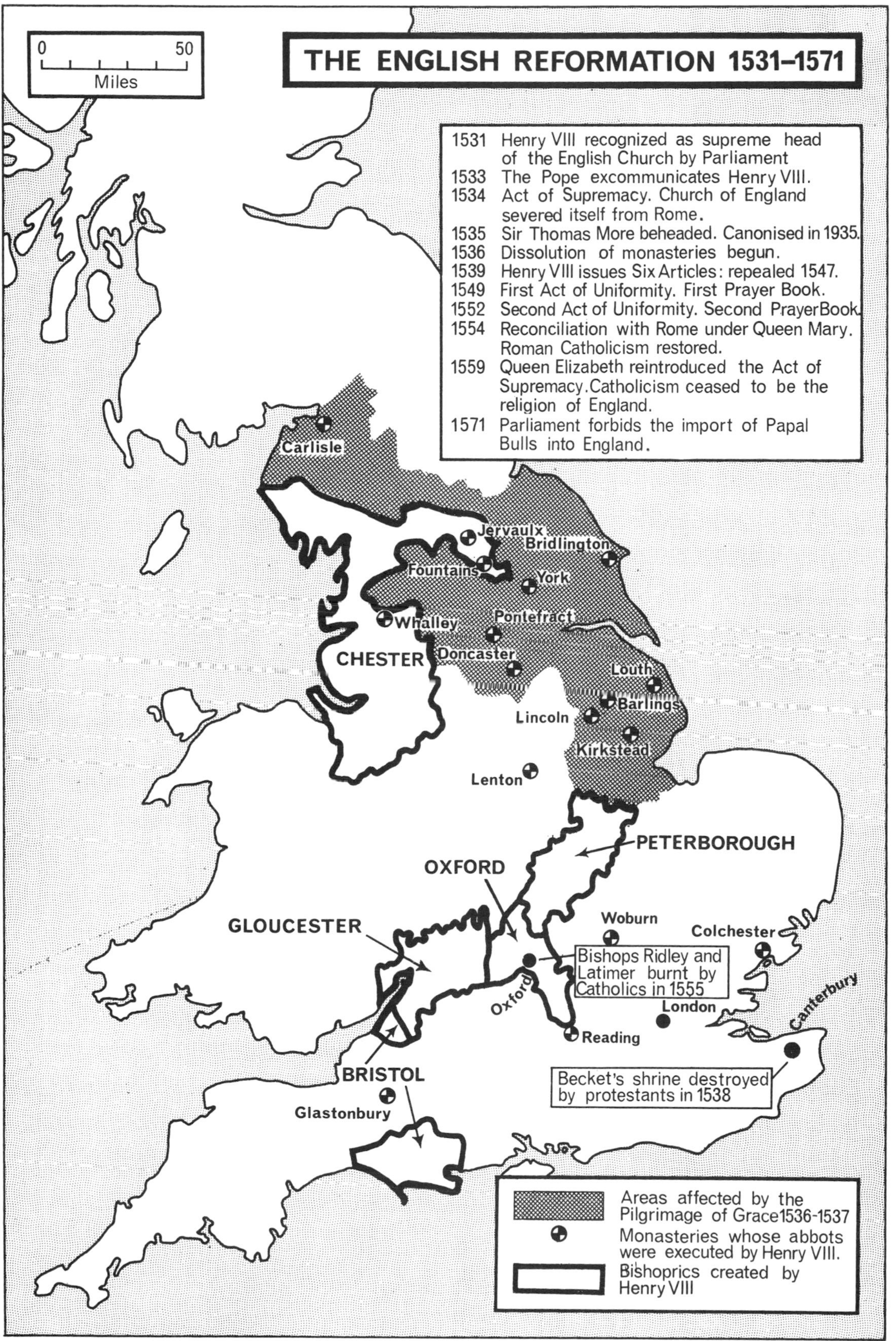
THE ENGLISH REFORMATION 1531–1571
0
50
Miles
1531 Henry VIII recognized as supreme head of the English Church by Parliament
1533 The Pope excommunicates Henry VIII.
1534 Act of Supremacy. Church of England severed itself from Rome.
1535 Sir Thomas More beheaded. Canonised in 1935.
1536 Dissolution of monasteries begun.
1539 Henry VIII issues Six Articles: repealed 1547.
1549 First Act of Uniformity. First Prayer Book.
1552 Second Act of Uniformity. Second PrayerBook.
1554 Reconciliation with Rome under Queen Mary. Roman Catholicism restored.
1559 Queen Elizabeth reintroduced the Act of Supremacy. Catholicism ceased to be the religion of England.
1571 Parliament forbids the import of Papal Bulls into England.
Carlisle
Jervaulx
Bridlington
Fountains
York
Whalley
Pontefract
CHESTER
Doncaster
Louth
Barlings
Lincoln
Kirkstead
Lenton
PETERBOROUGH
OXFORD
GLOUCESTER
Woburn
Colchester
Bishops Ridley and Latimer burnt by Catholics in 1555
Oxford
London
Canterbury
Reading
BRISTOL
Glastonbury
Becket's shrine destroyed by protestants in 1538
Areas affected by the Pilgrimage of Grace 1536-1537
Monasteries whose abbots were executed by Henry VIII.
Bishoprics created by Henry VIII

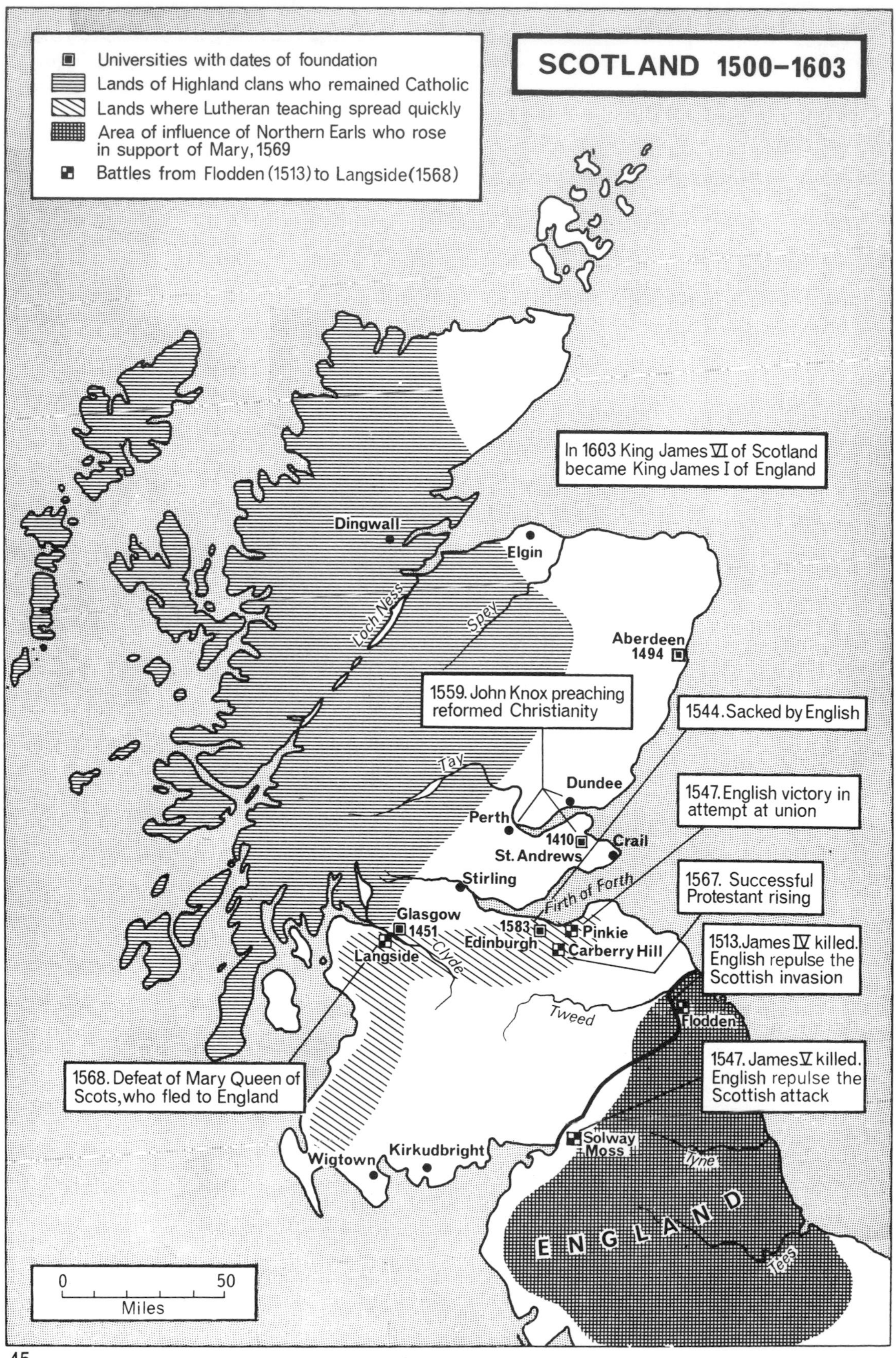
SCOTLAND 1500–1603
Universities with dates of foundation
Lands of Highland clans who remained Catholic
Lands where Lutheran teaching spread quickly
Area of influence of Northern Earls who rose in support of Mary, 1569
Battles from Flodden (1513) to Langside (1568)
In 1603 King James VI of Scotland became King James I of England
Dingwall
Elgin
Loch Ness
Spey
Aberdeen 1494
1559. John Knox preaching reformed Christianity
1544. Sacked by English
Tay
Dundee
Perth
1410
Crail
St. Andrews
1547. English victory in attempt at union
Stirling
Firth of Forth
1567. Successful Protestant rising
Glasgow 1451
1583
Pinkie
Edinburgh
Carberry Hill
Langside
Clyde
1513. James IV killed. English repulse the Scottish invasion
Tweed
Flodden
1547. James V killed. English repulse the Scottish attack
1568. Defeat of Mary Queen of Scots, who fled to England
Solway Moss
Tyne
Wigtown
Kirkudbright
ENGLAND
Tees
0
50
Miles

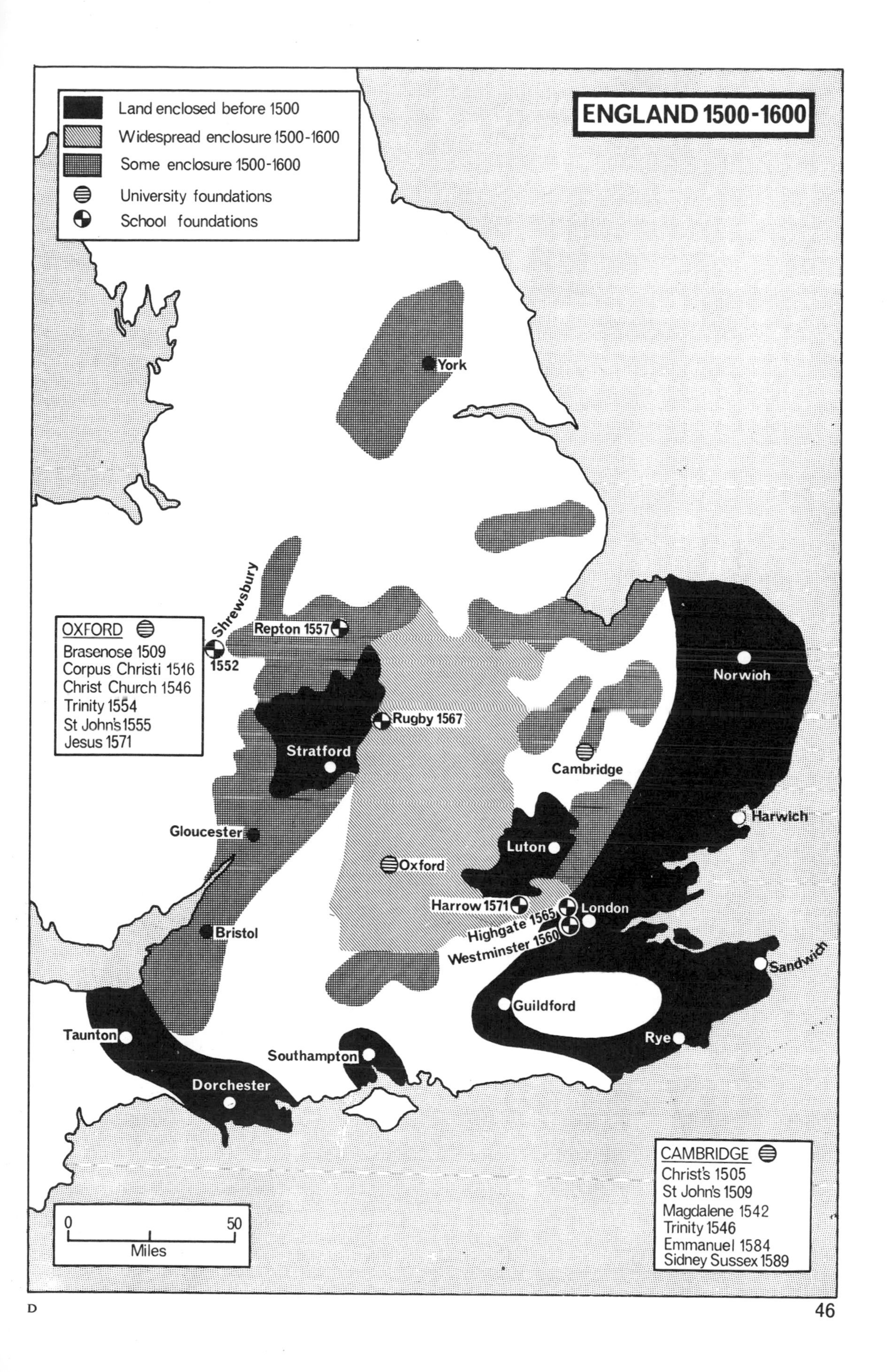
ENGLAND 1500-1600
Land enclosed before 1500
Widespread enclosure 1500-1600
Some enclosure 1500-1600
University foundations
School foundations
York
Shrewsbury
1552
Repton 1557
OXFORD
Brasenose 1509
Corpus Christi 1516
Christ Church 1546
Trinity 1554
St John's 1555
Jesus 1571
Rugby 1567
Stratford
Norwioh
Cambridge
Harwich
Gloucester
Luton
Oxford
Harrow 1571
London
Highgate 1565
Westminster 1560
Bristol
Sandwich
Guildford
Taunton
Southampton
Rye
Dorchester
CAMBRIDGE
Christ's 1505
St John's 1509
Magdalene 1542
Trinity 1546
Emmanuel 1584
Sidney Sussex 1589
0
50
Miles

D

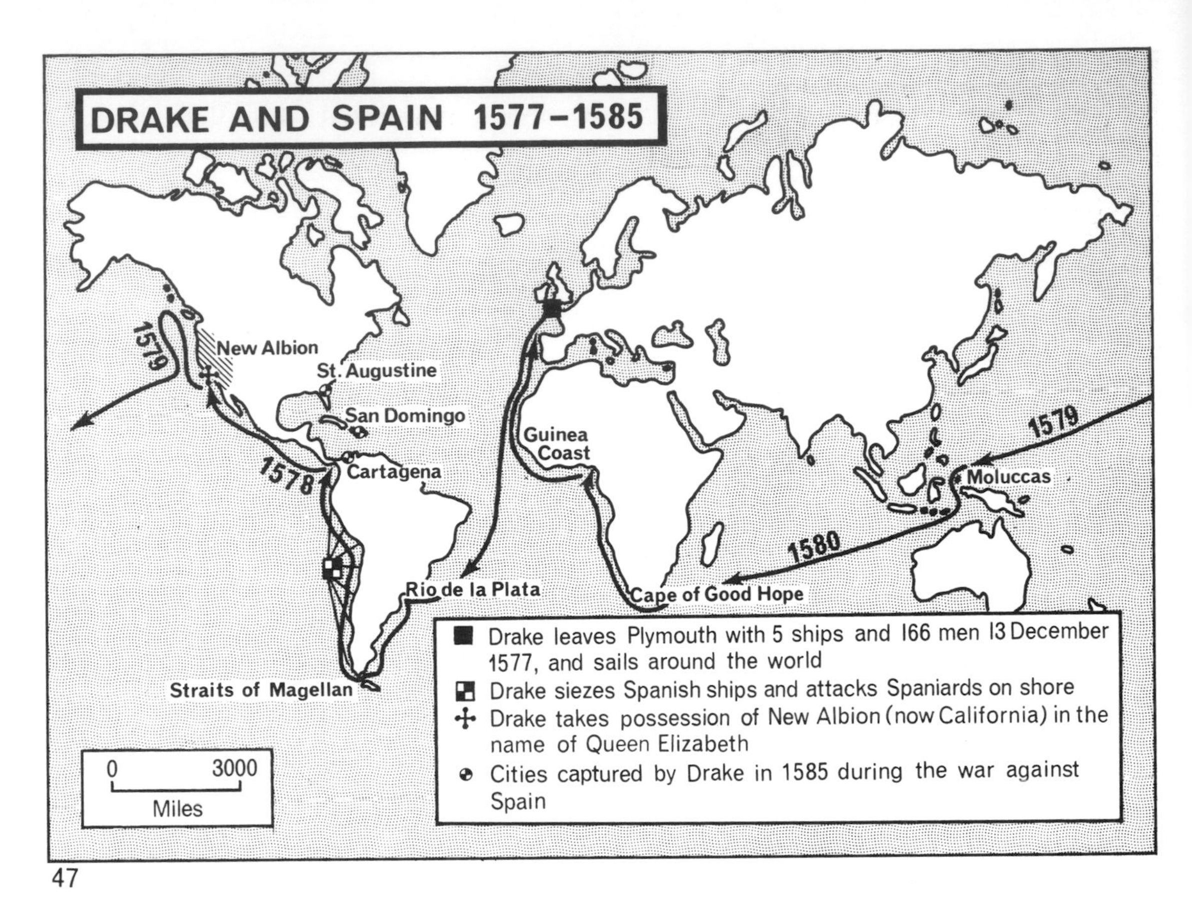
DRAKE AND SPAIN 1577–1585
1579
New Albion
St. Augustine
San Domingo
1578
Cartagena
Guinea Coast
Rio de la Plata
Cape of Good Hope
1579
Moluccas
1580
Straits of Magellan
Drake leaves Plymouth with 5 ships and 166 men 13 December 1577, and sails around the world
Drake siezes Spanish ships and attacks Spaniards on shore
Drake takes possession of New Albion (now California) in the name of Queen Elizabeth
Cities captured by Drake in 1585 during the war against Spain
0 3000
Miles

47

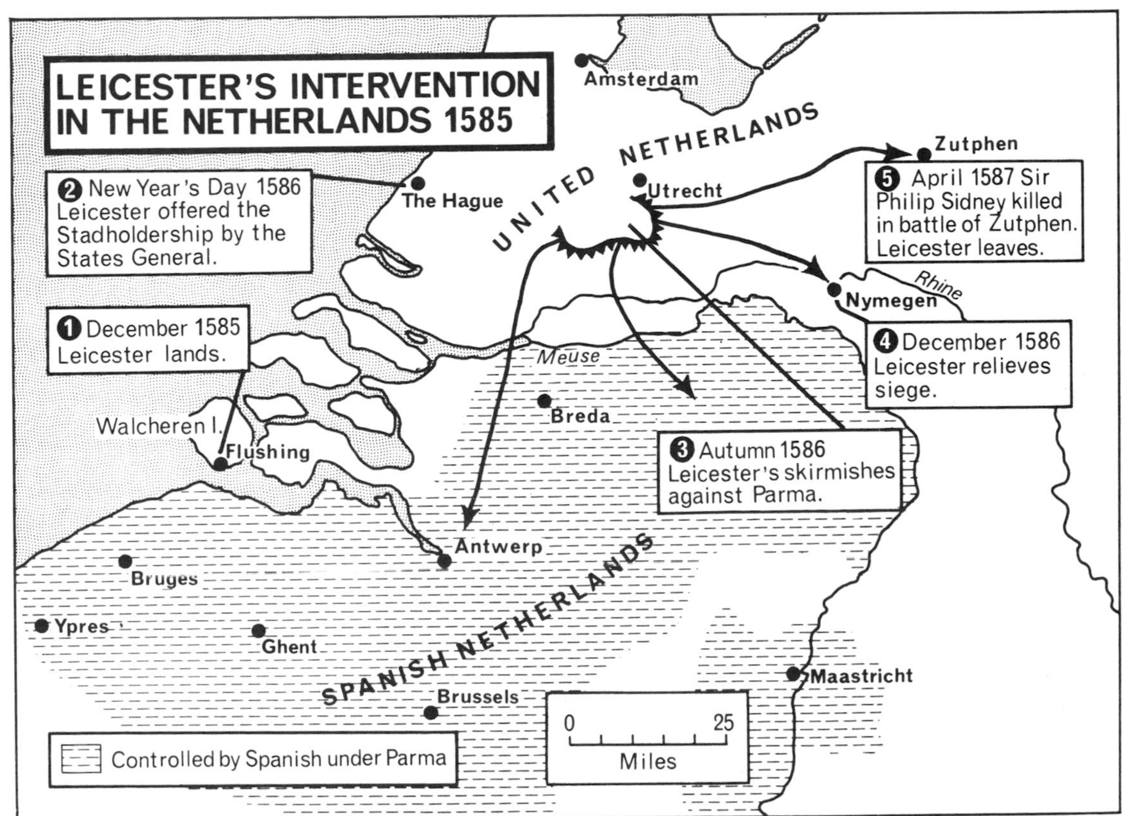
LEICESTER'S INTERVENTION IN THE NETHERLANDS 1585
Amsterdam
UNITED NETHERLANDS
Zutphen
Utrecht
The Hague
2 New Year's Day 1586 Leicester offered the Stadholdership by the States General.
5 April 1587 Sir Philip Sidney killed in battle of Zutphen. Leicester leaves.
Rhine
Nymegen
1 December 1585 Leicester lands.
4 December 1586 Leicester relieves siege.
Meuse
Breda
Walcheren I.
Flushing
3 Autumn 1586 Leicester's skirmishes against Parma.
Antwerp
Bruges
Ypres
Ghent
SPANISH NETHERLANDS
Maastricht
Brussels
Controlled by Spanish under Parma
0 25
Miles

48

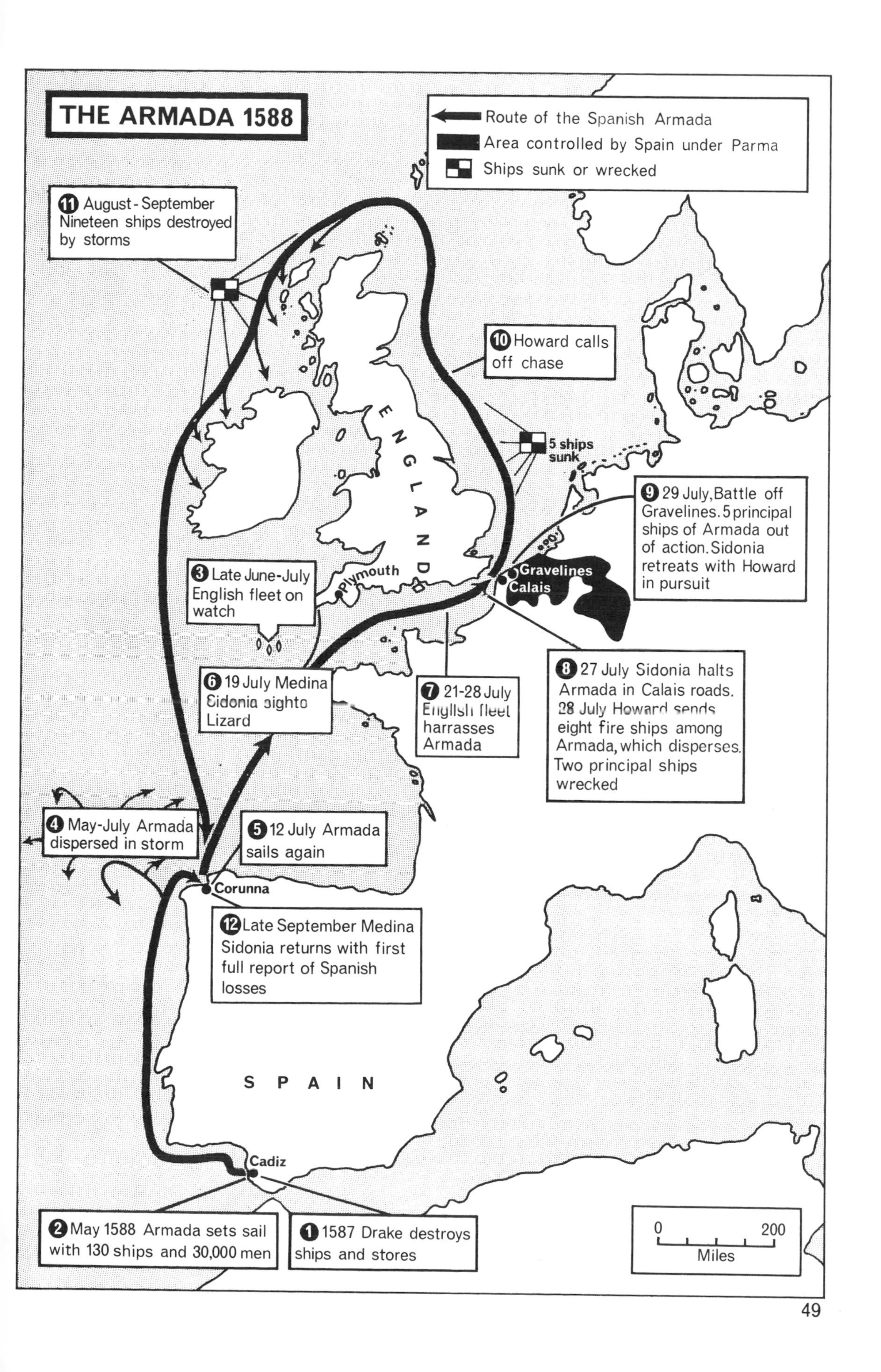
THE ARMADA 1588
Route of the Spanish Armada
Area controlled by Spain under Parma
Ships sunk or wrecked
11 August - September Nineteen ships destroyed by storms
10 Howard calls off chase
5 ships sunk
ENGLAND
9 29 July, Battle off Gravelines. 5 principal ships of Armada out of action. Sidonia retreats with Howard in pursuit
Gravelines
Calais
Plymouth
3 Late June-July English fleet on watch
6 19 July Medina Sidonia sights Lizard
7 21-28 July English fleet harrasses Armada
8 27 July Sidonia halts Armada in Calais roads. 28 July Howard sends eight fire ships among Armada, which disperses. Two principal ships wrecked
4 May-July Armada dispersed in storm
5 12 July Armada sails again
Corunna
12 Late September Medina Sidonia returns with first full report of Spanish losses
SPAIN
Cadiz
2 May 1588 Armada sets sail with 130 ships and 30,000 men
1 1587 Drake destroys ships and stores
0
200
Miles

50

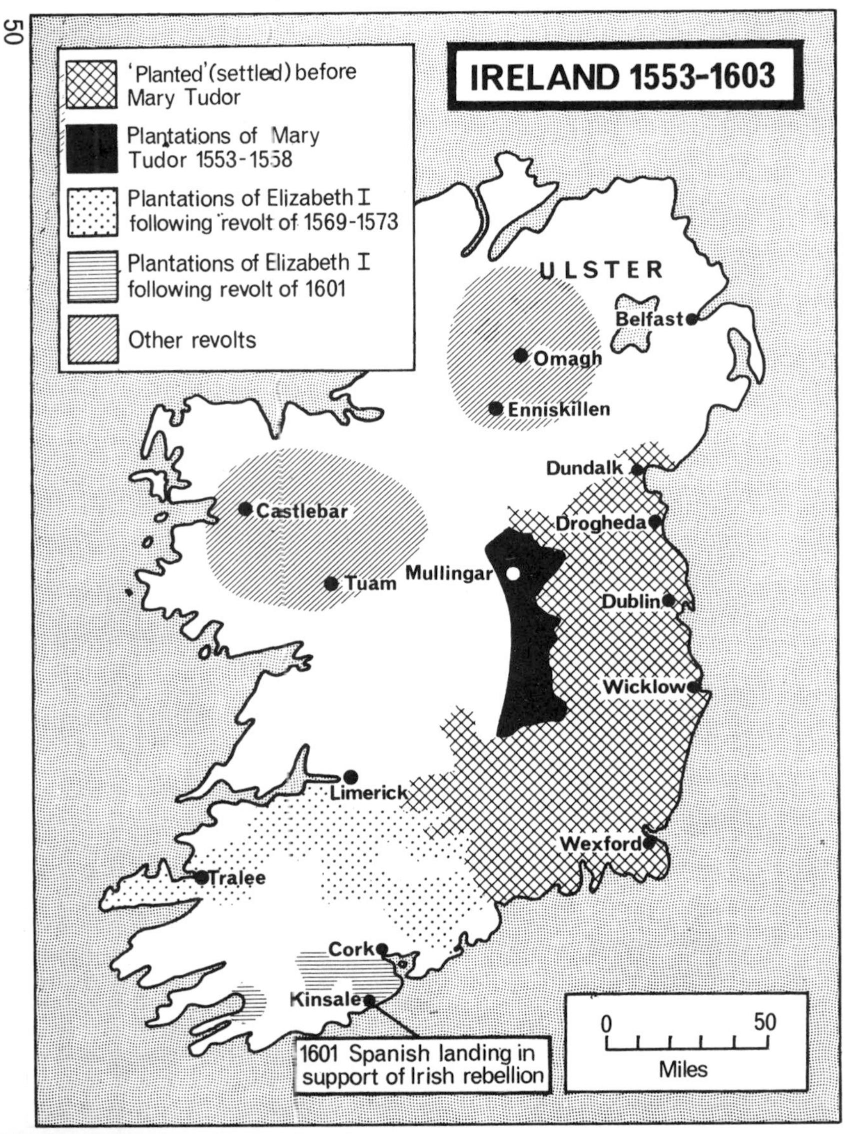

IRELAND 1553-1603
'Planted' (settled) before Mary Tudor
Plantations of Mary Tudor 1553-1558
Plantations of Elizabeth I following revolt of 1569-1573
Plantations of Elizabeth I following revolt of 1601
Other revolts
ULSTER
Belfast
Omagh
Enniskillen
Dundalk
Castlebar
Drogheda
Tuam
Mullingar
Dublin
Wicklow
Limerick
Wexford
Tralee
Cork
Kinsale
1601 Spanish landing in support of Irish rebellion
0
50
Miles

51

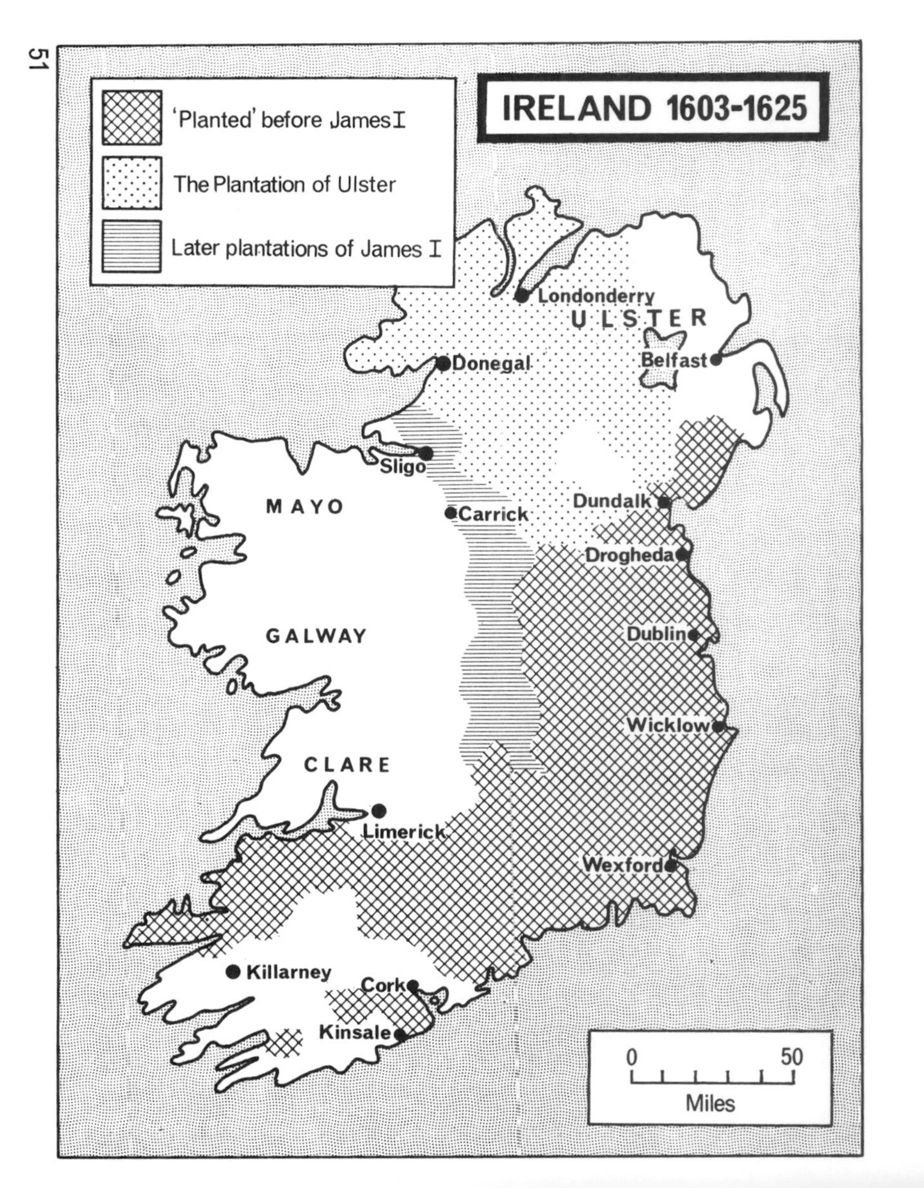

IRELAND 1603-1625
'Planted' before James I
The Plantation of Ulster
Later plantations of James I
Londonderry
ULSTER
Donegal
Belfast
Sligo
MAYO
Carrick
Dundalk
Drogheda
GALWAY
Dublin
Wicklow
CLARE
Limerick
Wexford
Killarney
Cork
Kinsale
0
50
Miles

THE CARIBBEAN 1562–1717

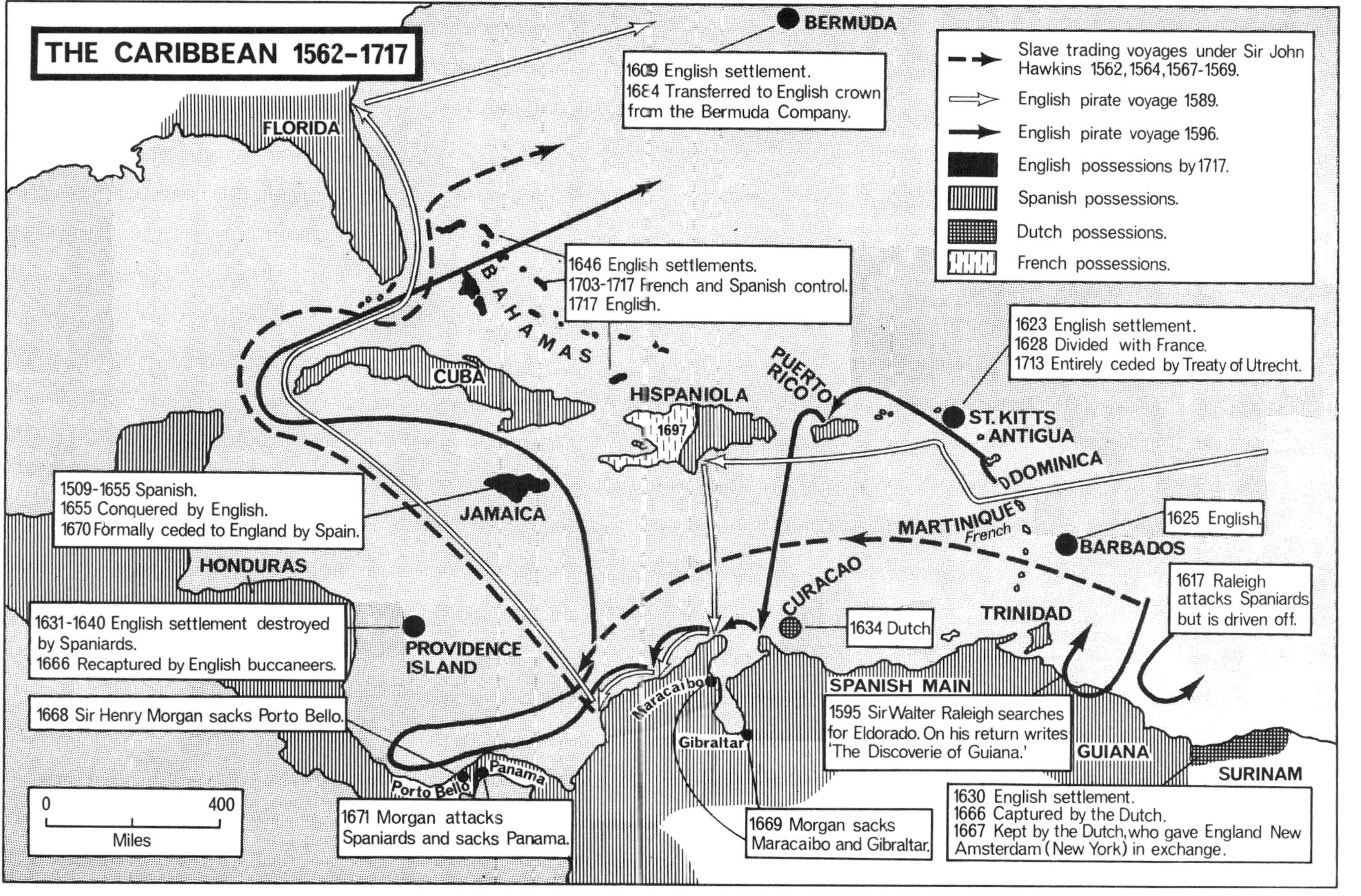

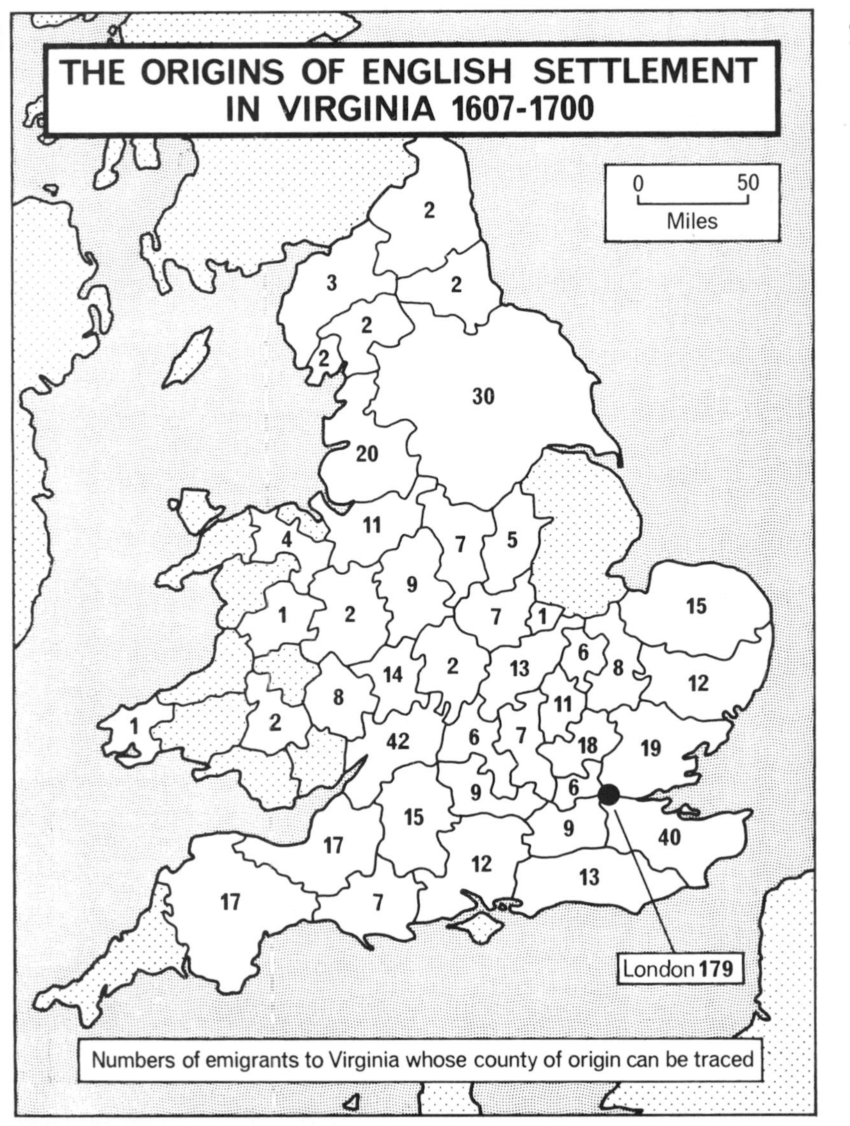
THE ORIGINS OF ENGLISH SETTLEMENT
IN VIRGINIA 1607-1700
0 50
Miles
London 179
Numbers of emigrants to Virginia whose county of origin can be traced

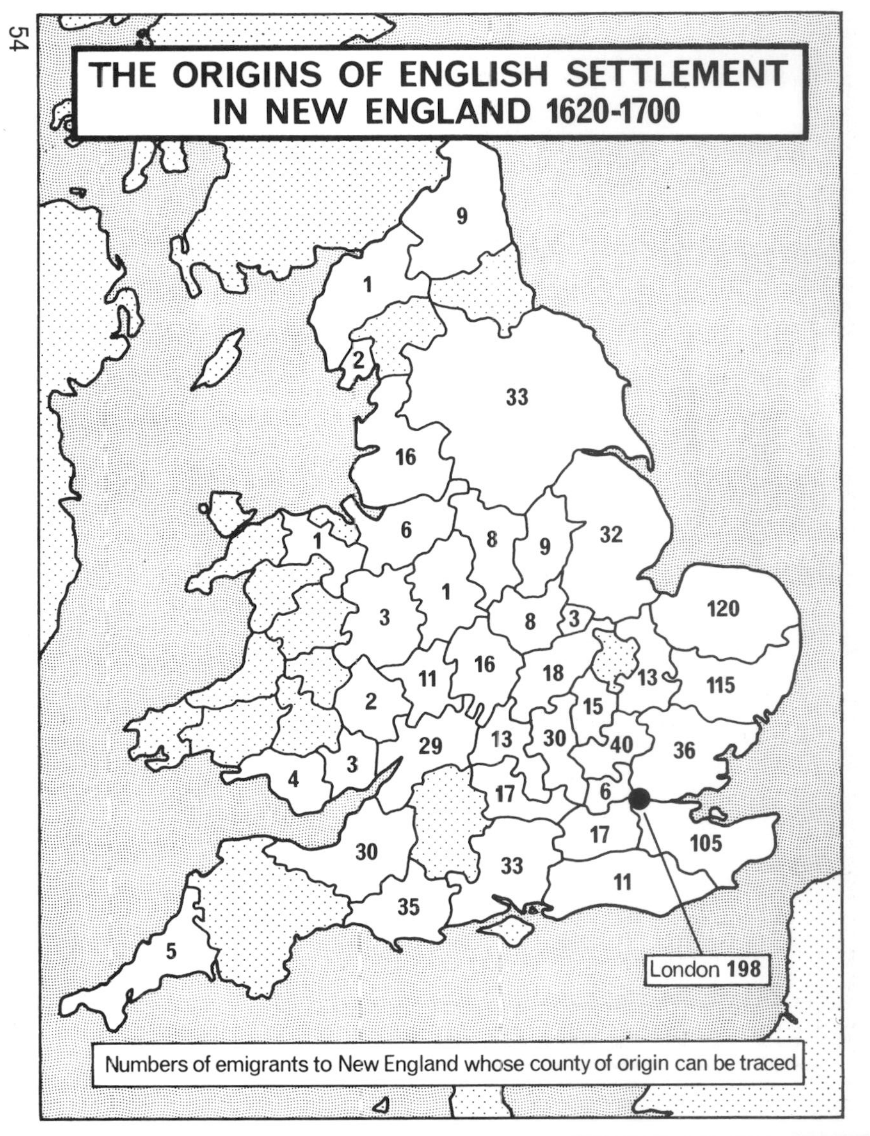
THE ORIGINS OF ENGLISH SETTLEMENT
IN NEW ENGLAND 1620-1700
London 198
Numbers of emigrants to New England whose county of origin can be traced

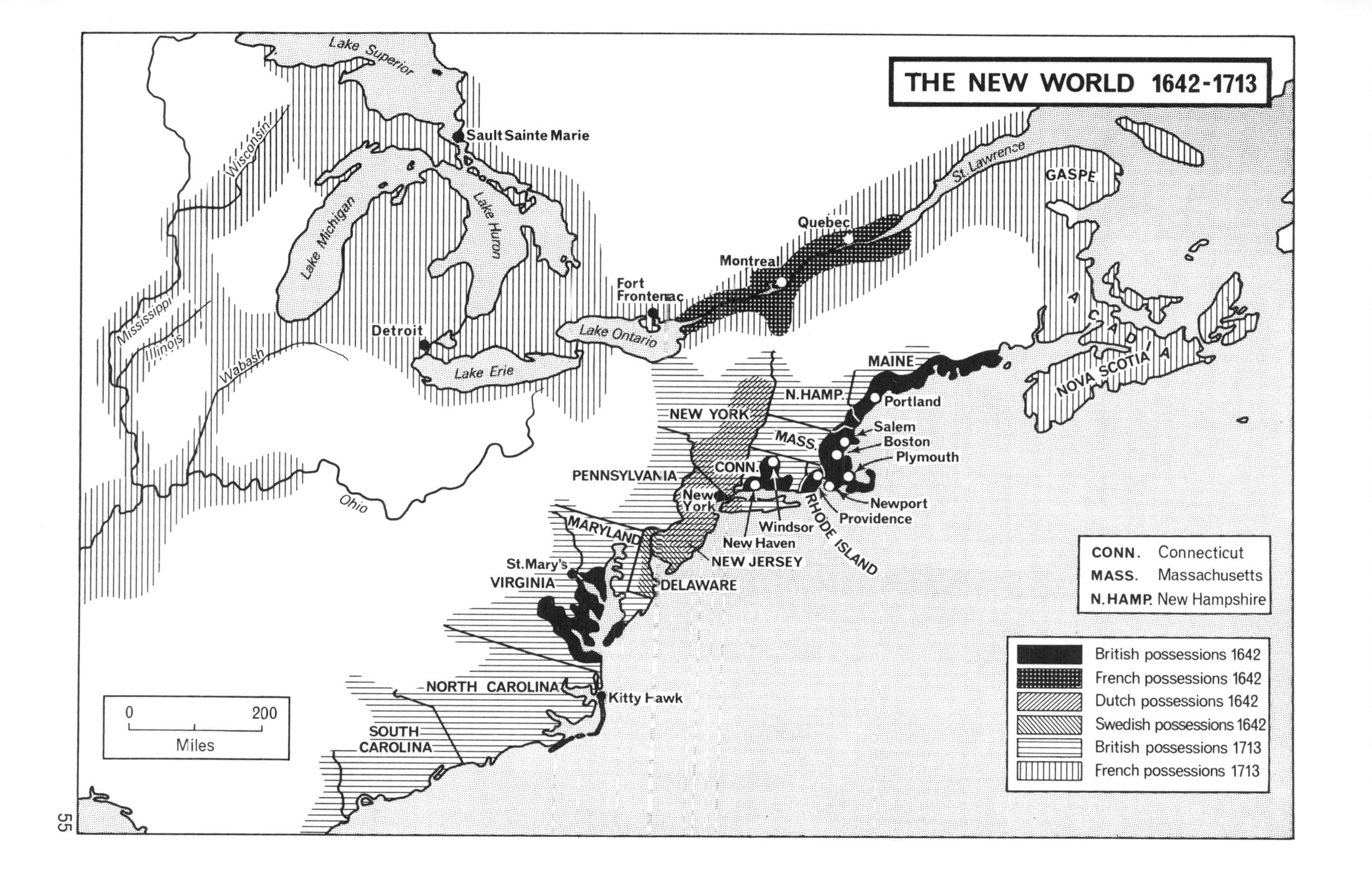
THE NEW WORLD 1642-1713
Lake Superior
Sault Sainte Marie
Wisconsin
Lake Michigan
Lake Huron
Mississippi
Illinois
Wabash
Detroit
Lake Erie
Ohio
Fort Frontenac
Lake Ontario
Montreal
Quebec
St. Lawrence
GASPE
ACADIA
NOVA SCOTIA
MAINE
N.HAMP.
Portland
Salem
Boston
Plymouth
Newport
Providence
RHODE ISLAND
MASS.
CONN.
Windsor
New Haven
NEW YORK
New York
NEW JERSEY
PENNSYLVANIA
DELAWARE
MARYLAND
St. Mary's
VIRGINIA
NORTH CAROLINA
Kitty Hawk
SOUTH CAROLINA
0
200
Miles
CONN. Connecticut
MASS. Massachusetts
N. HAMP. New Hampshire
British possessions 1642
French possessions 1642
Dutch possessions 1642
Swedish possessions 1642
British possessions 1713
French possessions 1713

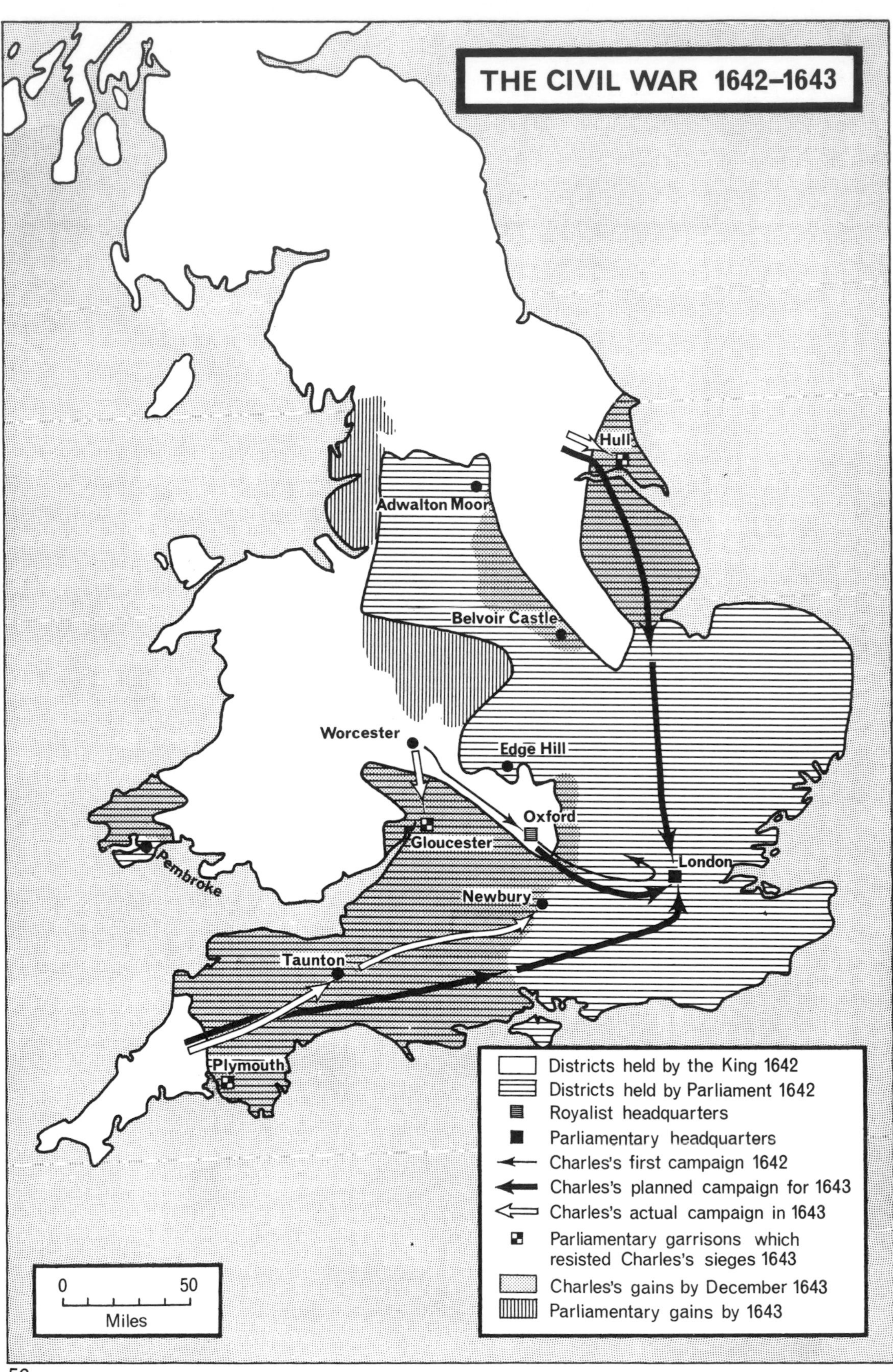
THE CIVIL WAR 1642–1643
Hull
Adwalton Moor
Belvoir Castle
Worcester
Edge Hill
Oxford
Gloucester
Pembroke
London
Newbury
Taunton
Plymouth
Districts held by the King 1642
Districts held by Parliament 1642
Royalist headquarters
Parliamentary headquarters
Charles's first campaign 1642
Charles's planned campaign for 1643
Charles's actual campaign in 1643
Parliamentary garrisons which resisted Charles's sieges 1643
Charles's gains by December 1643
Parliamentary gains by 1643
0
50
Miles

THE CIVIL WAR 1644–1646
In May 1646 King Charles surrendered to the Scottish Army at Newark. In February 1647 the Scots sold the King to Parliament for £400,000. He was beheaded on 30 January 1649.
Carlisle
Marston Moor
Hull
Preston
Bolton
Liverpool
Stockport
Sandal Castle
Hulme
Newark
Nantwich
Belvoir Castle
Shrewsbury
Ashby
Lichfield
Naseby
Holmby House
Banbury
Cropredy Bridge
Gloucester
Oxford
Donnington Castle
Bridgewater
Taunton
Lyme Regis
Corfe Castle
Plymouth
0
50
Miles
The Eastern Association: main recruiting ground for Parliamentary Army 1643
Campaign of Prince Rupert to Marston Moor.
Parliamentary advances to Marston Moor, where the Royalists were defeated 2 July 1644
Area controlled by Parliament in December 1644.
Area gained by Parliament by December 1645.
Districts held by the King in May 1646.
Area gained by Parliament by December 1646.

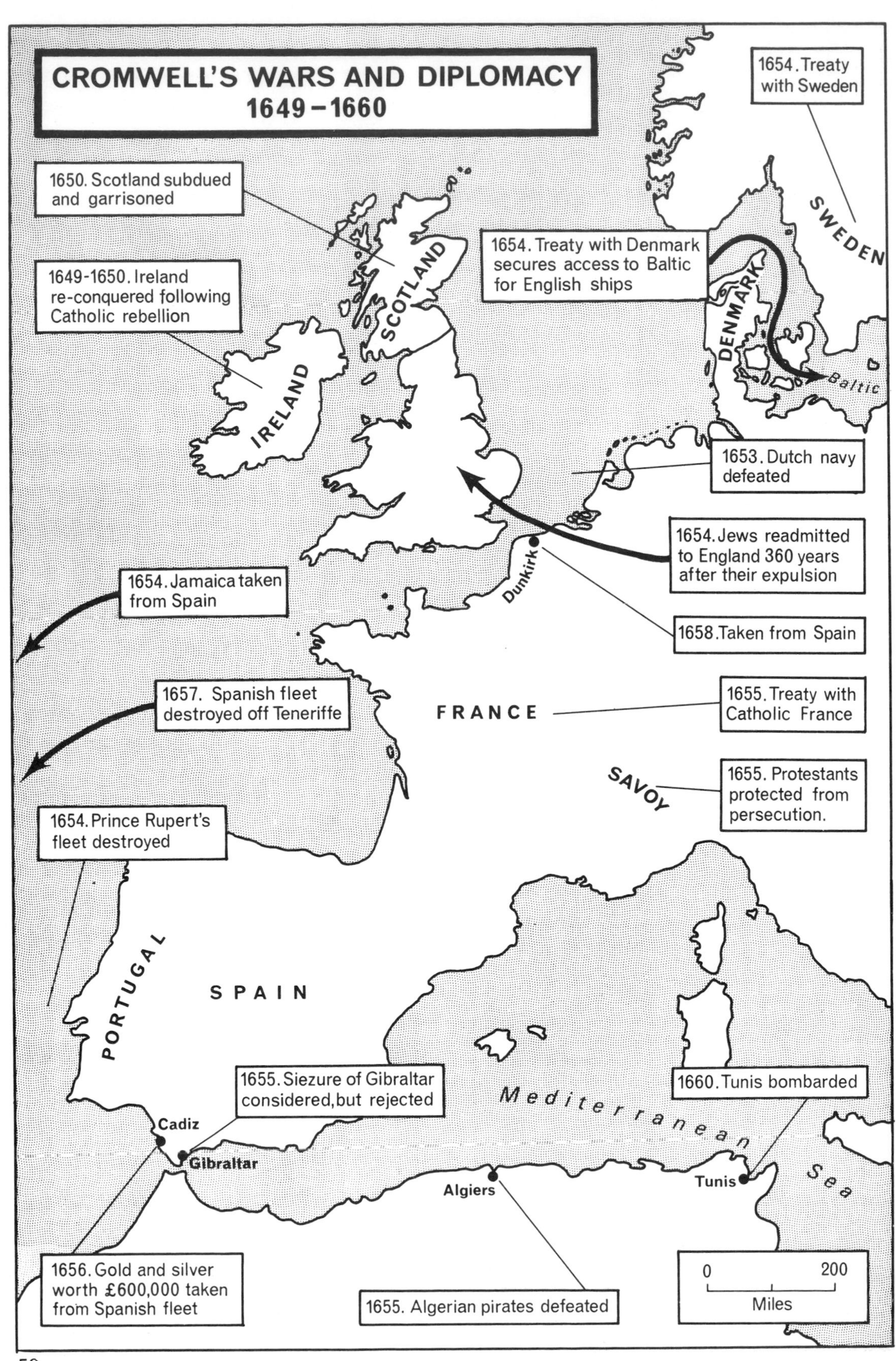

CROMWELL'S WARS AND DIPLOMACY
1649–1660
1654. Treaty with Sweden
1650. Scotland subdued and garrisoned
1654. Treaty with Denmark secures access to Baltic for English ships
1649-1650. Ireland re-conquered following Catholic rebellion
SWEDEN
SCOTLAND
DENMARK
IRELAND
Baltic
1653. Dutch navy defeated
1654. Jews readmitted to England 360 years after their expulsion
Dunkirk
1654. Jamaica taken from Spain
1658. Taken from Spain
1657. Spanish fleet destroyed off Teneriffe
FRANCE
1655. Treaty with Catholic France
SAVOY
1655. Protestants protected from persecution.
1654. Prince Rupert's fleet destroyed
PORTUGAL
SPAIN
1655. Siezure of Gibraltar considered, but rejected
1660. Tunis bombarded
Mediterranean Sea
Cadiz
Gibraltar
Algiers
Tunis
1656. Gold and silver worth £600,000 taken from Spanish fleet
1655. Algerian pirates defeated
0
200
Miles

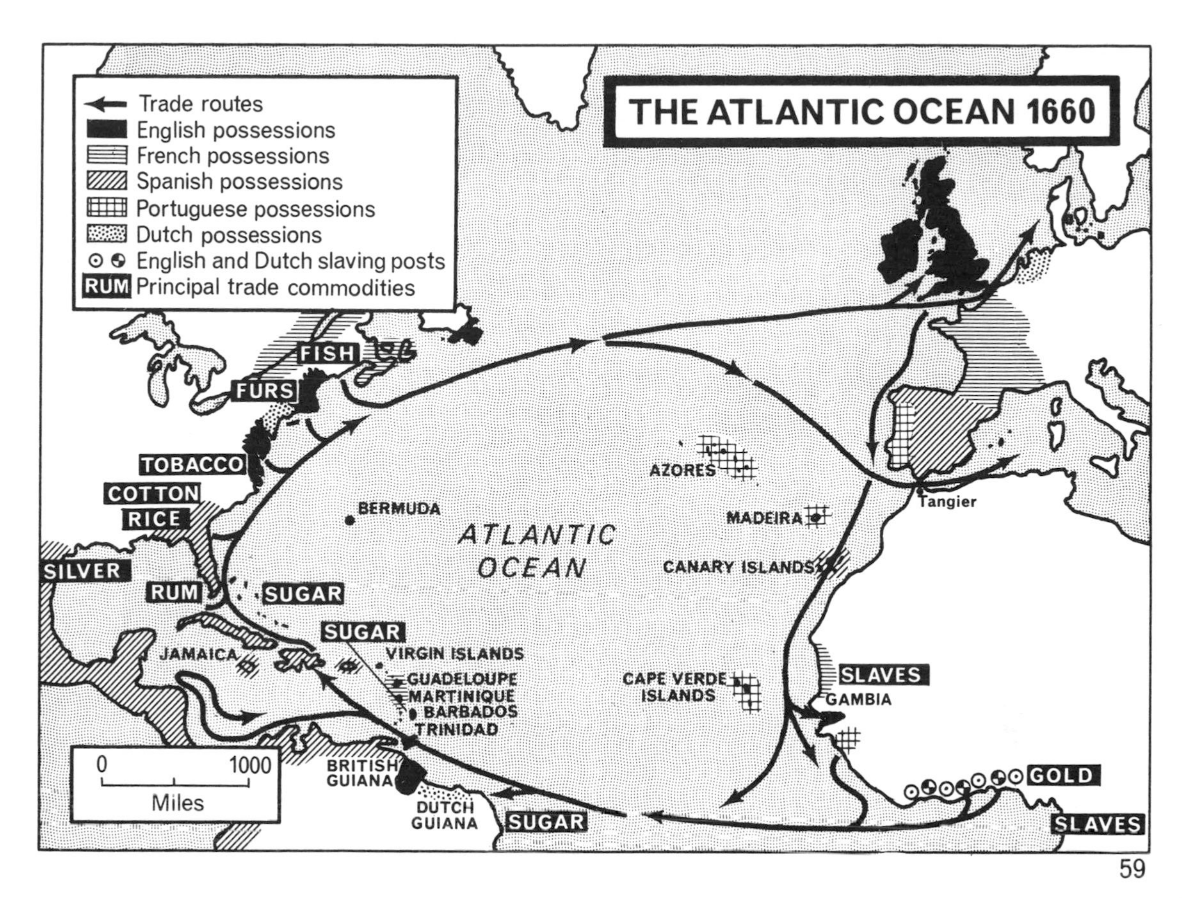
THE ATLANTIC OCEAN 1660
Trade routes
English possessions
French possessions
Spanish possessions
Portuguese possessions
Dutch possessions
English and Dutch slaving posts
RUM Principal trade commodities
FISH
FURS
TOBACCO
COTTON
RICE
SILVER
RUM
SUGAR
SUGAR
JAMAICA
VIRGIN ISLANDS
GUADELOUPE
MARTINIQUE
BARBADOS
TRINIDAD
BRITISH GUIANA
DUTCH GUIANA
SUGAR
BERMUDA
ATLANTIC OCEAN
AZORES
MADEIRA
CANARY ISLANDS
CAPE VERDE ISLANDS
Tangier
SLAVES
GAMBIA
GOLD
SLAVES
0 1000 Miles

59

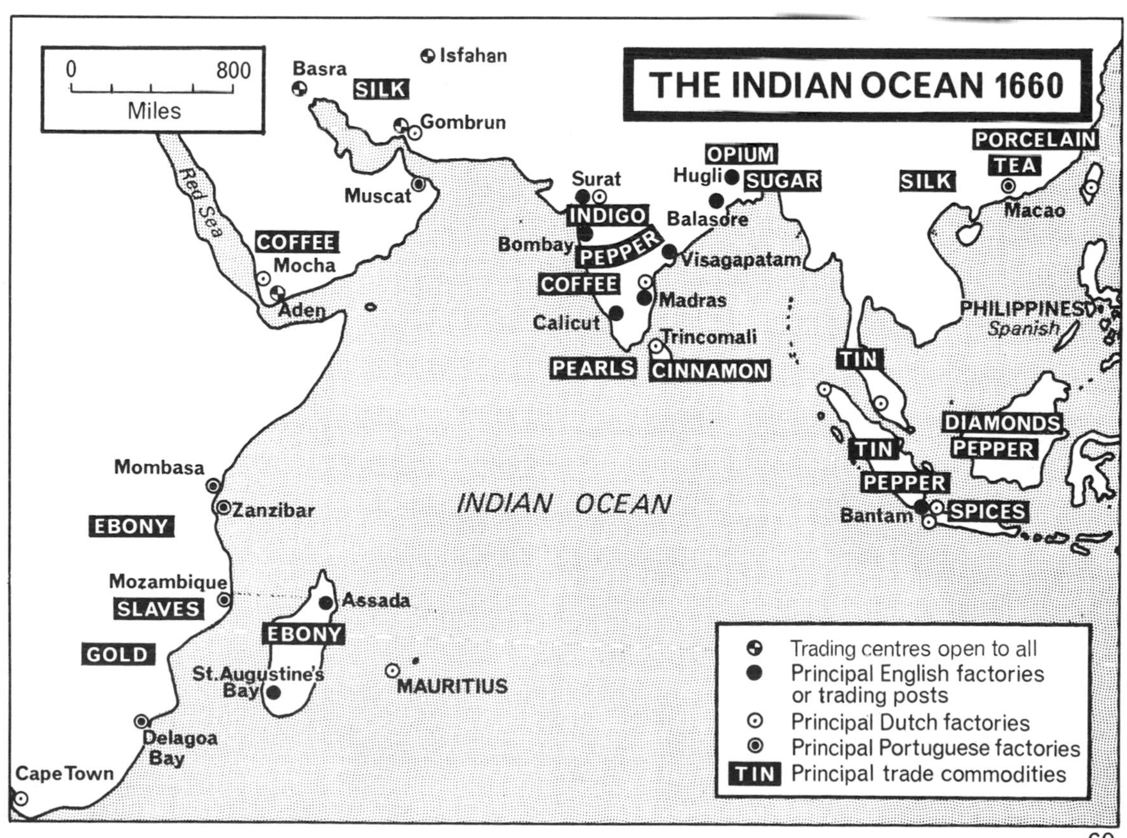
THE INDIAN OCEAN 1660
0 800 Miles
Basra
SILK
Isfahan
Gombrun
Muscat
Red Sea
COFFEE
Mocha
Aden
Surat
INDIGO
Bombay
PEPPER
COFFEE
Calicut
PEARLS
OPIUM
Hugli
SUGAR
Balasore
Visagapatam
Madras
Trincomali
CINNAMON
SILK
PORCELAIN
TEA
Macao
PHILIPPINES
Spanish
TIN
TIN
DIAMONDS
PEPPER
PEPPER
Bantam
SPICES
INDIAN OCEAN
Mombasa
Zanzibar
EBONY
Mozambique
SLAVES
GOLD
Assada
EBONY
St. Augustine's Bay
MAURITIUS
Delagoa Bay
Cape Town
Trading centres open to all
Principal English factories or trading posts
Principal Dutch factories
Principal Portuguese factories
TIN Principal trade commodities

THE THREE DUTCH WARS

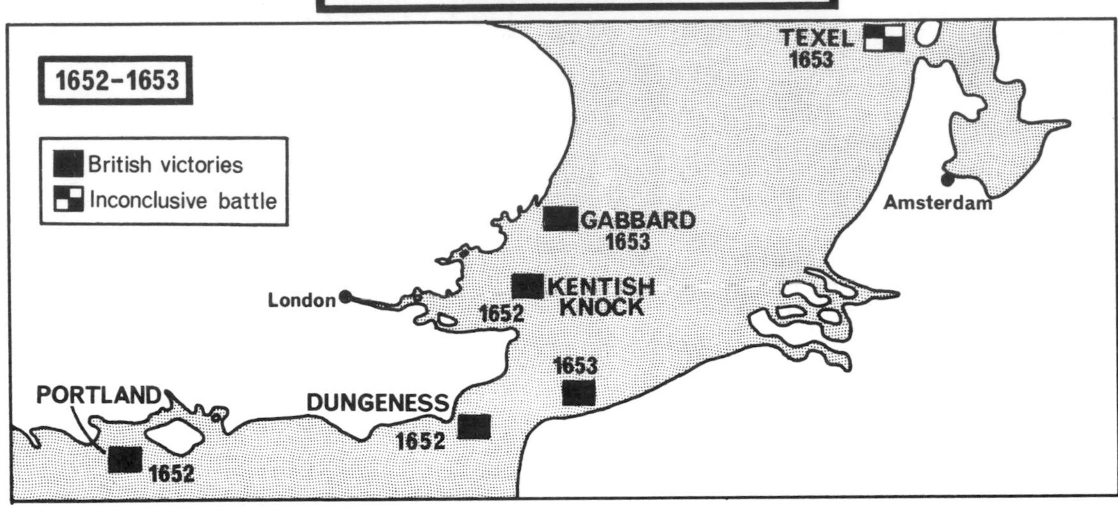

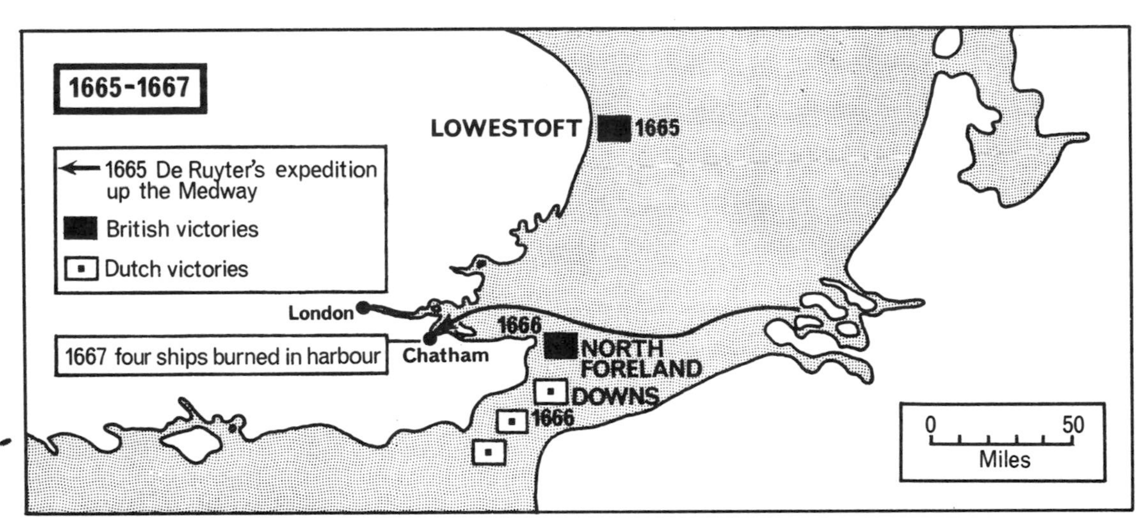

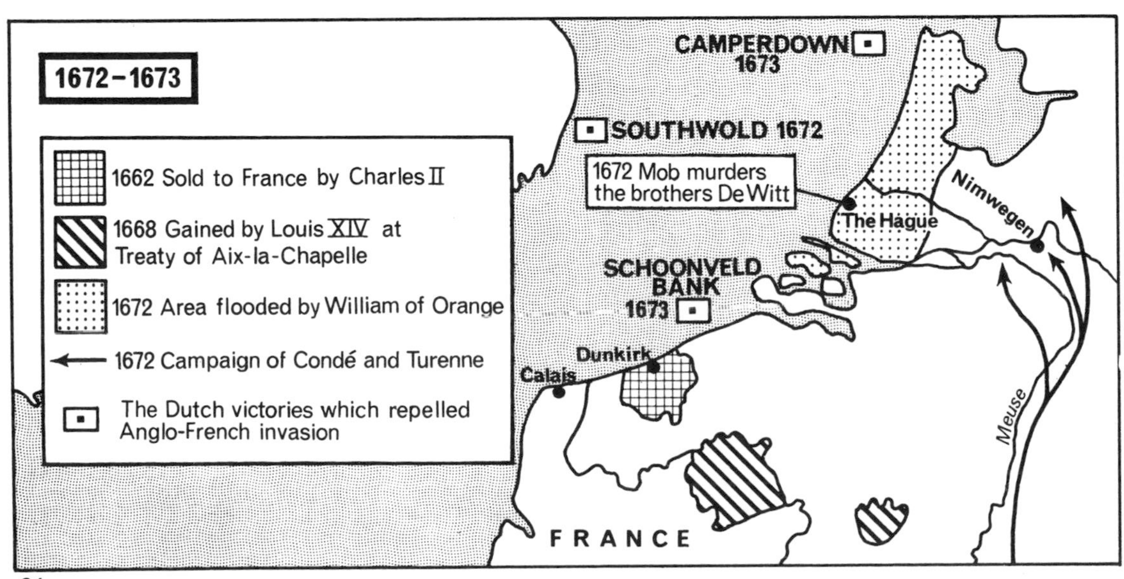

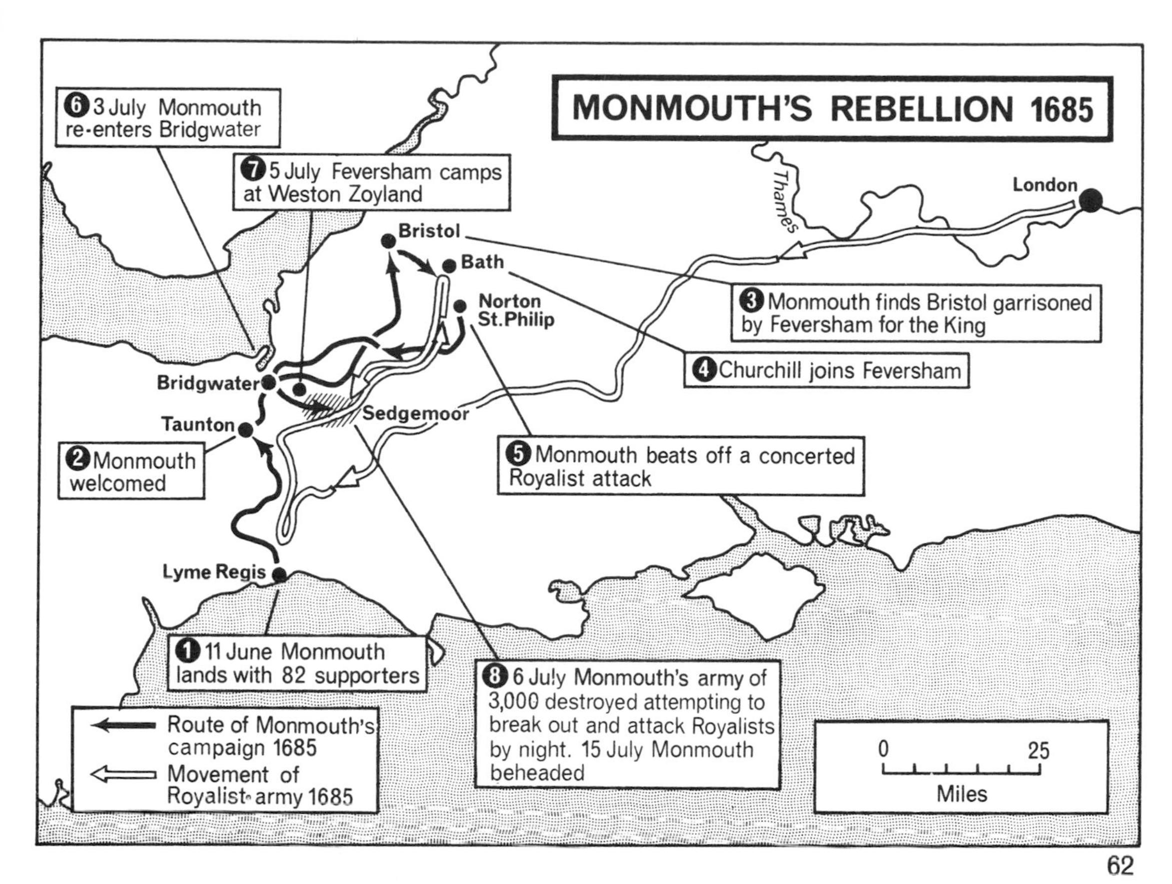

MONMOUTH'S REBELLION 1685
1 11 June Monmouth lands with 82 supporters
2 Monmouth welcomed
3 Monmouth finds Bristol garrisoned by Feversham for the King
4 Churchill joins Feversham
5 Monmouth beats off a concerted Royalist attack
6 3 July Monmouth re-enters Bridgwater
7 5 July Feversham camps at Weston Zoyland
8 6 July Monmouth's army of 3,000 destroyed attempting to break out and attack Royalists by night. 15 July Monmouth beheaded
Thames
London
Bristol
Bath
Norton St.Philip
Bridgwater
Taunton
Sedgemoor
Lyme Regis
Route of Monmouth's campaign 1685
Movement of Royalist army 1685
0
25
Miles

62

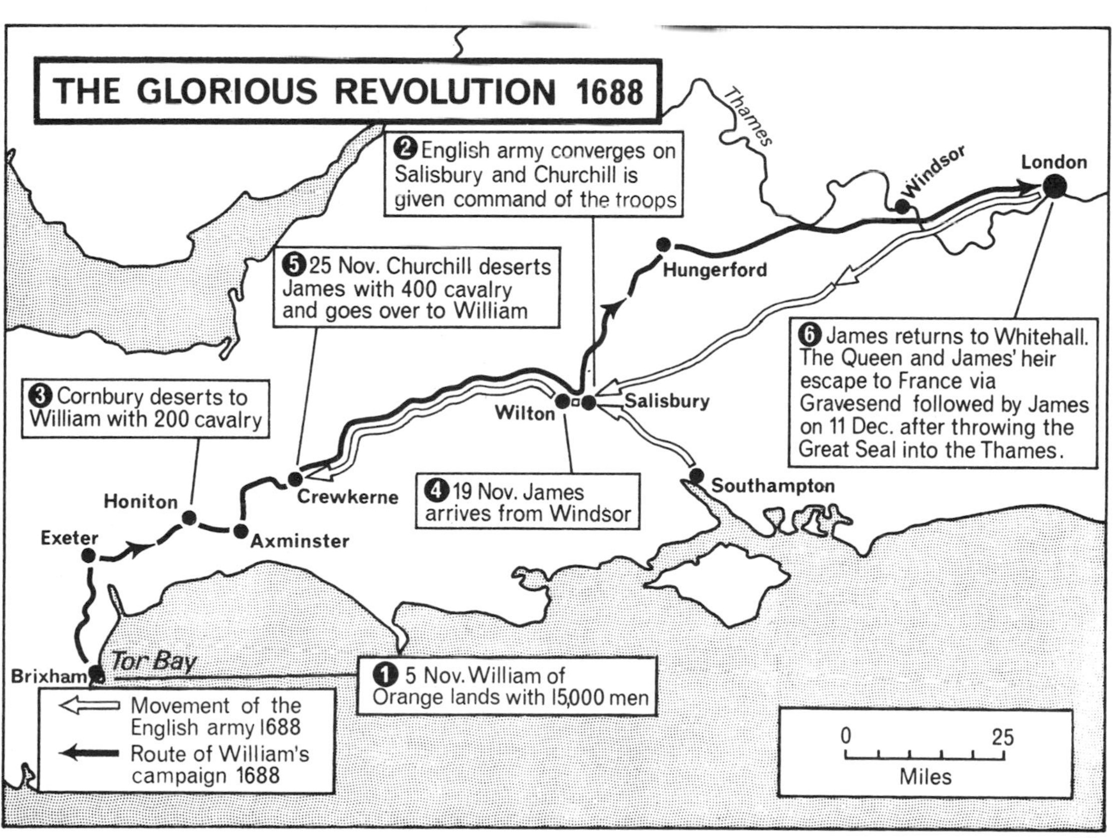

THE GLORIOUS REVOLUTION 1688
1 5 Nov. William of Orange lands with 15,000 men
2 English army converges on Salisbury and Churchill is given command of the troops
3 Cornbury deserts to William with 200 cavalry
4 19 Nov. James arrives from Windsor
5 25 Nov. Churchill deserts James with 400 cavalry and goes over to William
6 James returns to Whitehall. The Queen and James' heir escape to France via Gravesend followed by James on 11 Dec. after throwing the Great Seal into the Thames.
Thames
Windsor
London
Hungerford
Wilton
Salisbury
Southampton
Crewkerne
Honiton
Axminster
Exeter
Tor Bay
Brixham
Movement of the English army 1688
Route of William's campaign 1688
0
25
Miles

63

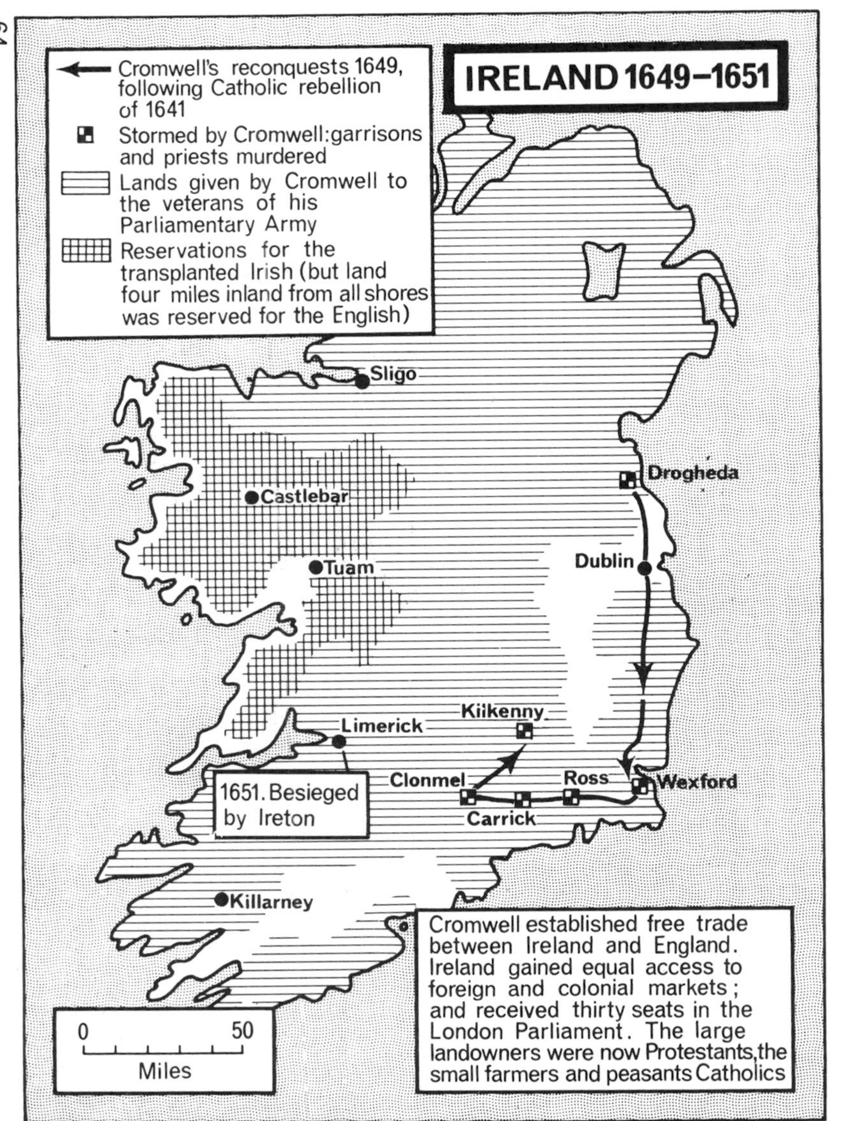
IRELAND 1649–1651
Cromwell's reconquests 1649, following Catholic rebellion of 1641
Stormed by Cromwell: garrisons and priests murdered
Lands given by Cromwell to the veterans of his Parliamentary Army
Reservations for the transplanted Irish (but land four miles inland from all shores was reserved for the English)
Sligo
Castlebar
Tuam
Drogheda
Dublin
Limerick
Kilkenny
Clonmel
Ross
Wexford
Carrick
1651. Besieged by Ireton
Killarney
0
50
Miles
Cromwell established free trade between Ireland and England. Ireland gained equal access to foreign and colonial markets; and received thirty seats in the London Parliament. The large landowners were now Protestants, the small farmers and peasants Catholics

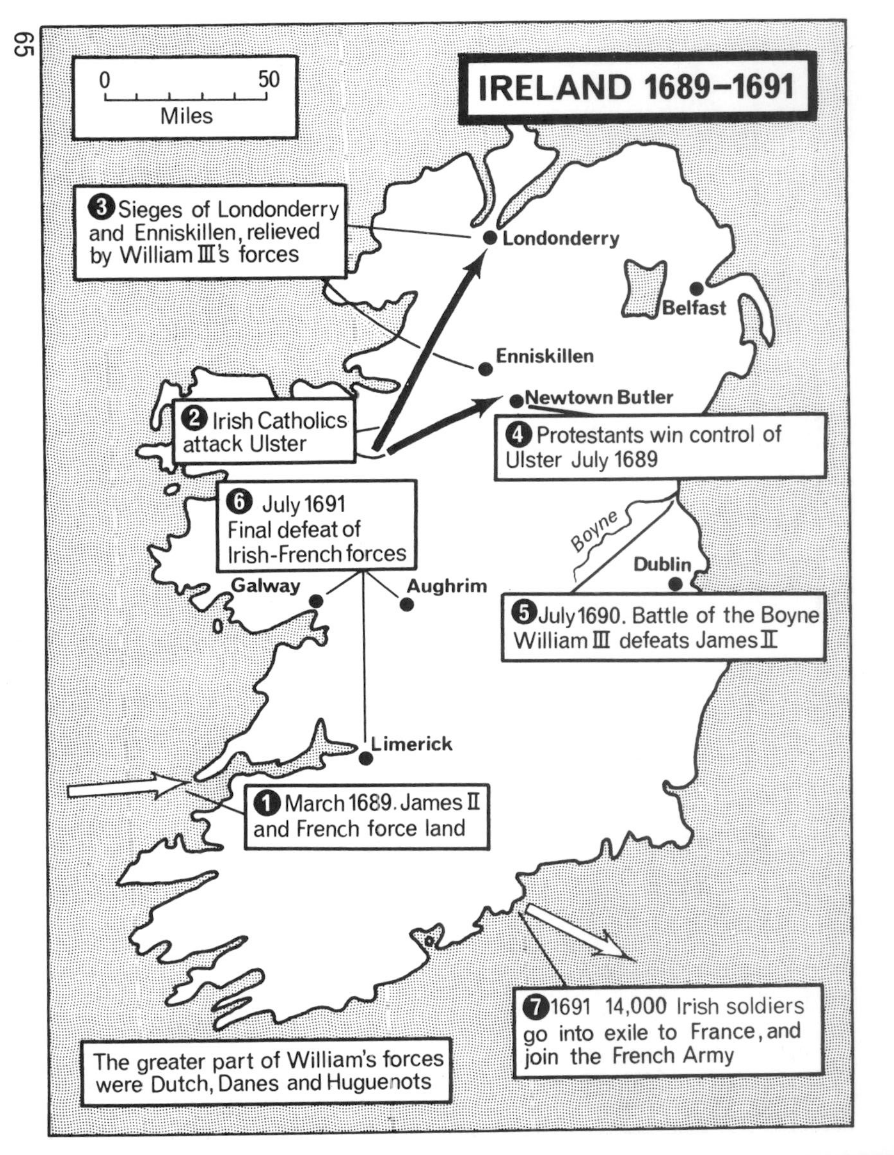
IRELAND 1689–1691
0
50
Miles
3 Sieges of Londonderry and Enniskillen, relieved by William III's forces
Londonderry
Belfast
Enniskillen
Newtown Butler
2 Irish Catholics attack Ulster
4 Protestants win control of Ulster July 1689
6 July 1691 Final defeat of Irish-French forces
Boyne
Dublin
Galway
Aughrim
5 July 1690. Battle of the Boyne William III defeats James II
Limerick
1 March 1689. James II and French force land
7 1691 14,000 Irish soldiers go into exile to France, and join the French Army
The greater part of William's forces were Dutch, Danes and Huguenots

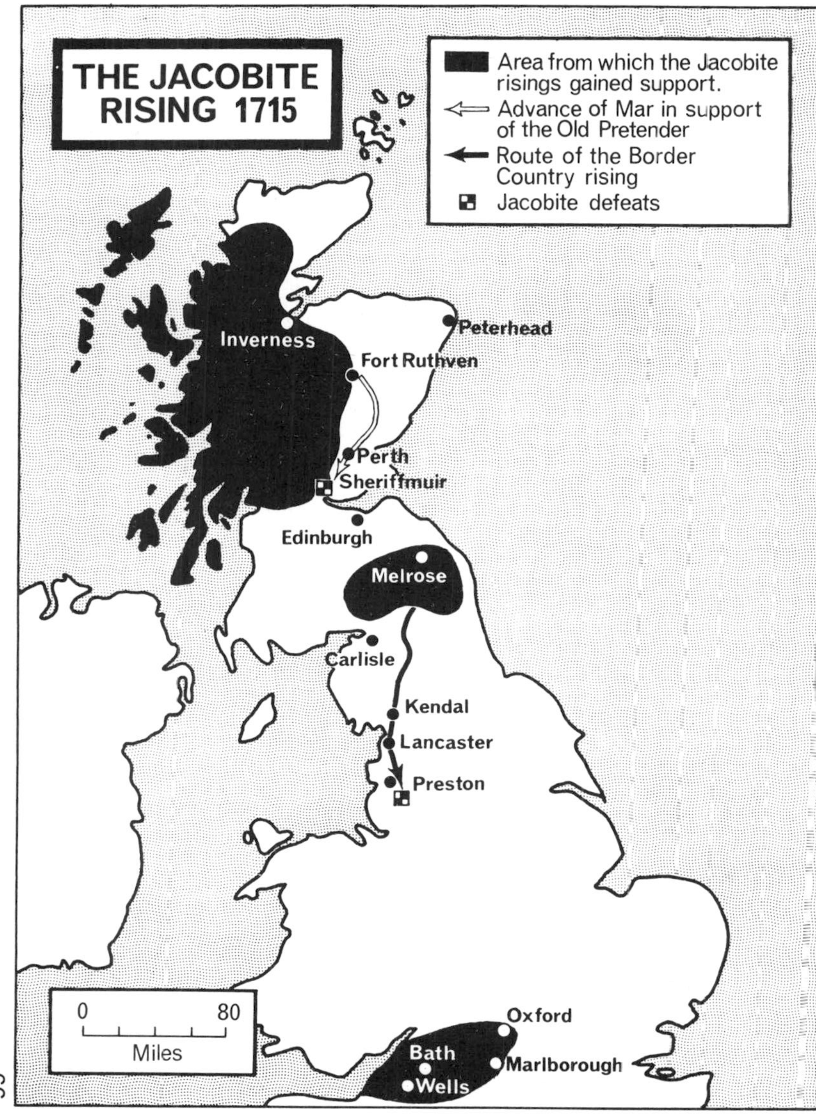
THE JACOBITE RISING 1715
Area from which the Jacobite risings gained support.
Advance of Mar in support of the Old Pretender
Route of the Border Country rising
Jacobite defeats
Inverness
Fort Ruthven
Peterhead
Perth
Sheriffmuir
Edinburgh
Melrose
Carlisle
Kendal
Lancaster
Preston
Oxford
Bath
Marlborough
Wells
0
80
Miles

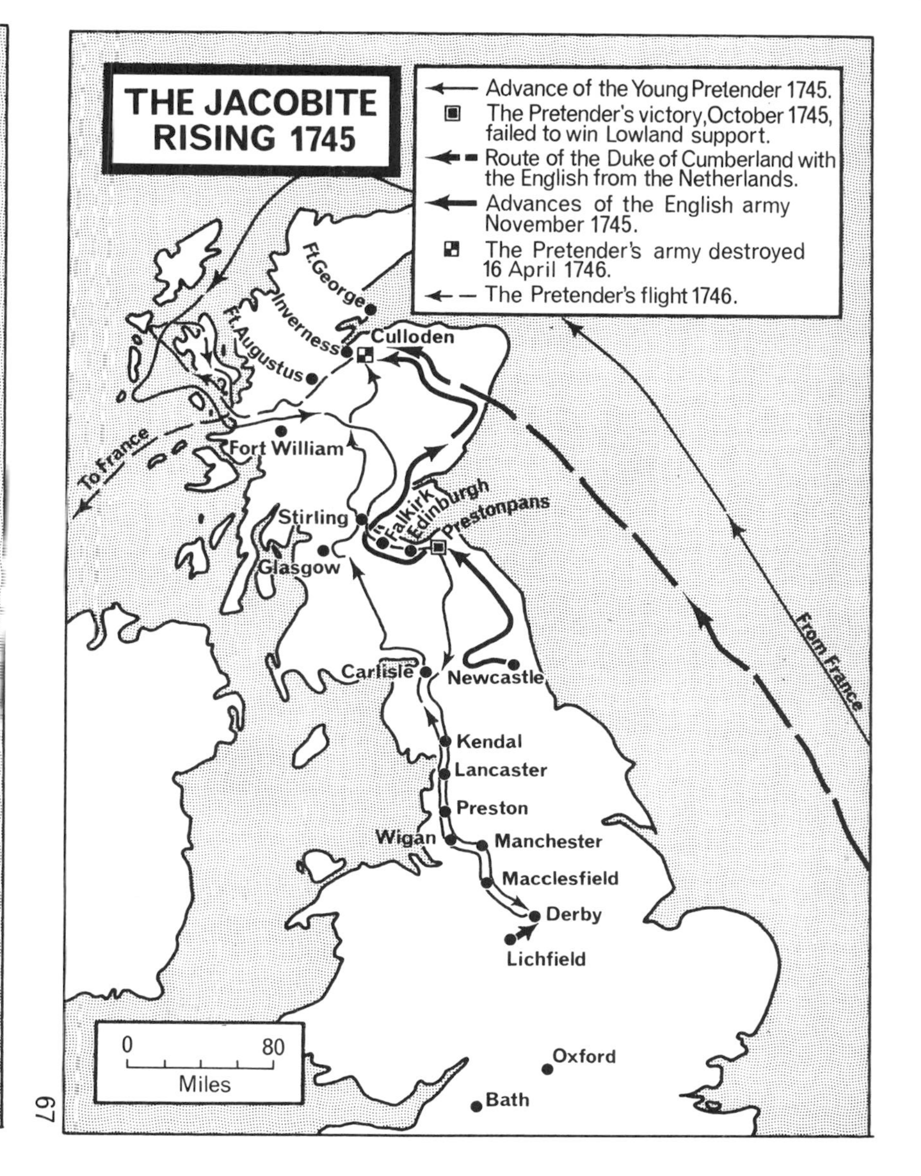
THE JACOBITE RISING 1745
Advance of the Young Pretender 1745.
The Pretender's victory, October 1745, failed to win Lowland support.
Route of the Duke of Cumberland with the English from the Netherlands.
Advances of the English army November 1745.
The Pretender's army destroyed 16 April 1746.
The Pretender's flight 1746.
Ft.George
Inverness
Culloden
Ft.Augustus
Fort William
To France
From France
Stirling
Falkirk
Edinburgh
Prestonpans
Glasgow
Carlisle
Newcastle
Kendal
Lancaster
Preston
Wigan
Manchester
Macclesfield
Derby
Lichfield
Oxford
Bath
0
80
Miles

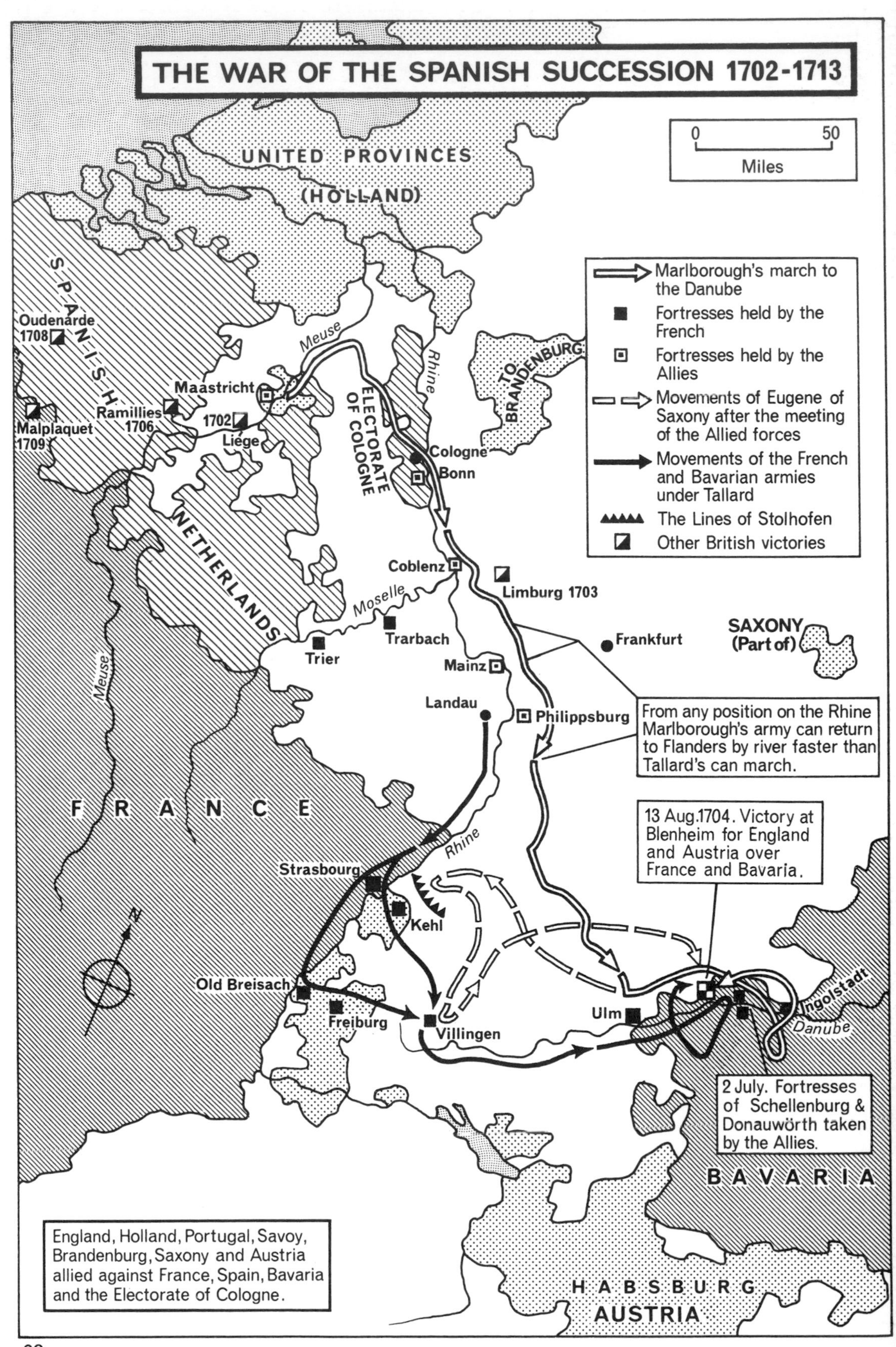
THE WAR OF THE SPANISH SUCCESSION 1702-1713
0
50
Miles
UNITED PROVINCES
(HOLLAND)
SPANISH
NETHERLANDS
ELECTORATE OF COLOGNE
TO BRANDENBURG
Marlborough's march to the Danube
Fortresses held by the French
Fortresses held by the Allies
Movements of Eugene of Saxony after the meeting of the Allied forces
Movements of the French and Bavarian armies under Tallard
The Lines of Stolhofen
Other British victories
Oudenarde 1708
Malplaquet 1709
Ramillies 1706
Maastricht
1702
Liège
Meuse
Rhine
Cologne
Bonn
Coblenz
Limburg 1703
Moselle
Trarbach
Trier
Frankfurt
SAXONY (Part of)
Mainz
Landau
Philippsburg
From any position on the Rhine Marlborough's army can return to Flanders by river faster than Tallard's can march.
FRANCE
13 Aug.1704. Victory at Blenheim for England and Austria over France and Bavaria.
Strasbourg
Kehl
Old Breisach
Freiburg
Villingen
Ulm
Ingolstadt
Danube
2 July. Fortresses of Schellenburg & Donauwörth taken by the Allies.
BAVARIA
England, Holland, Portugal, Savoy, Brandenburg, Saxony and Austria allied against France, Spain, Bavaria and the Electorate of Cologne.
HABSBURG
AUSTRIA
N

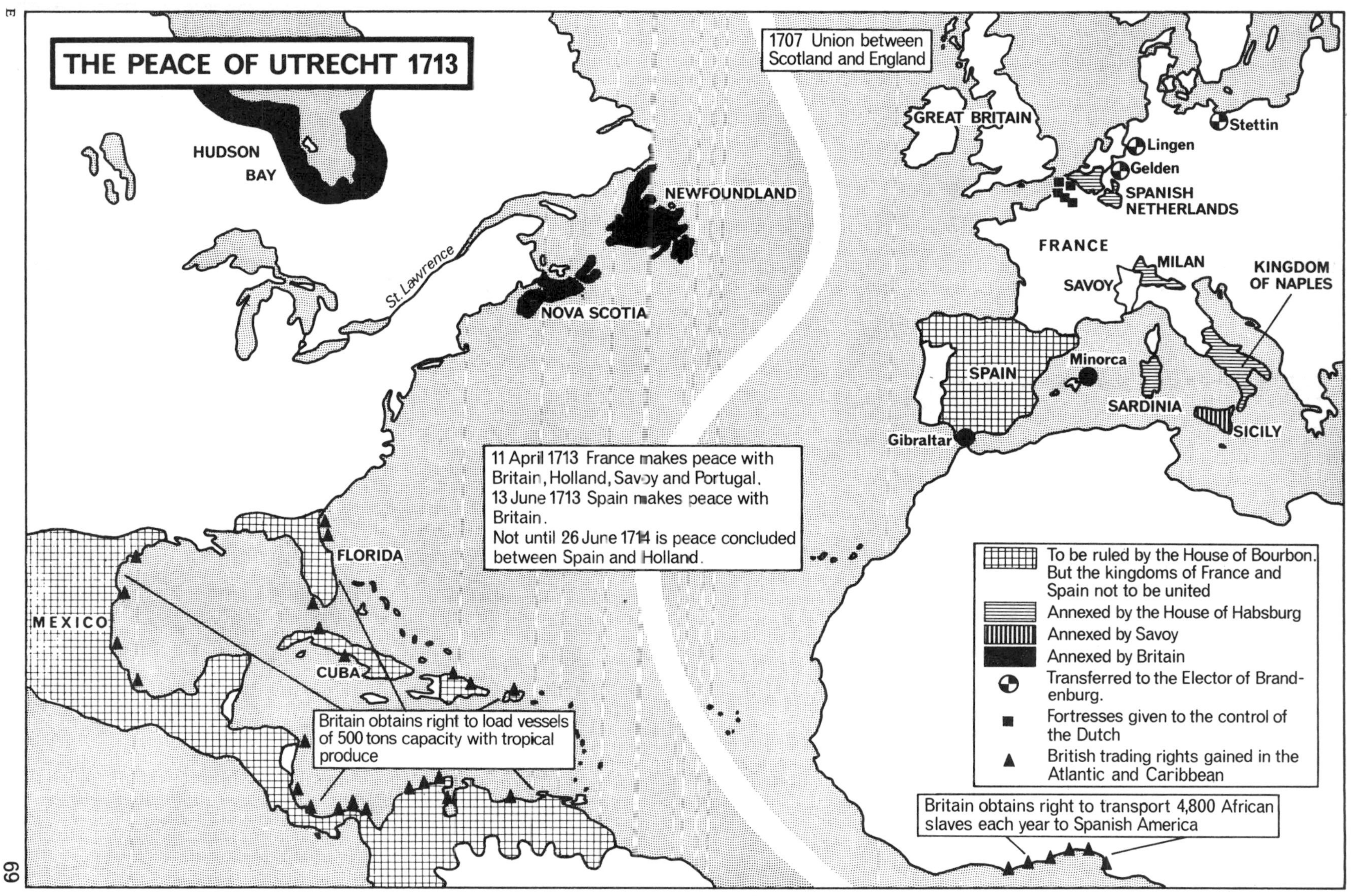

THE PEACE OF UTRECHT 1713
HUDSON BAY
St. Lawrence
NEWFOUNDLAND
NOVA SCOTIA
1707 Union between Scotland and England
GREAT BRITAIN
Stettin
Lingen
Gelden
SPANISH NETHERLANDS
FRANCE
SAVOY
MILAN
KINGDOM OF NAPLES
SPAIN
Minorca
SARDINIA
SICILY
Gibraltar
11 April 1713 France makes peace with Britain, Holland, Savoy and Portugal.
13 June 1713 Spain makes peace with Britain.
Not until 26 June 1714 is peace concluded between Spain and Holland.
FLORIDA
MEXICO
CUBA
Britain obtains right to load vessels of 500 tons capacity with tropical produce
To be ruled by the House of Bourbon. But the kingdoms of France and Spain not to be united
Annexed by the House of Habsburg
Annexed by Savoy
Annexed by Britain
Transferred to the Elector of Brandenburg.
Fortresses given to the control of the Dutch
British trading rights gained in the Atlantic and Caribbean
Britain obtains right to transport 4,800 African slaves each year to Spanish America

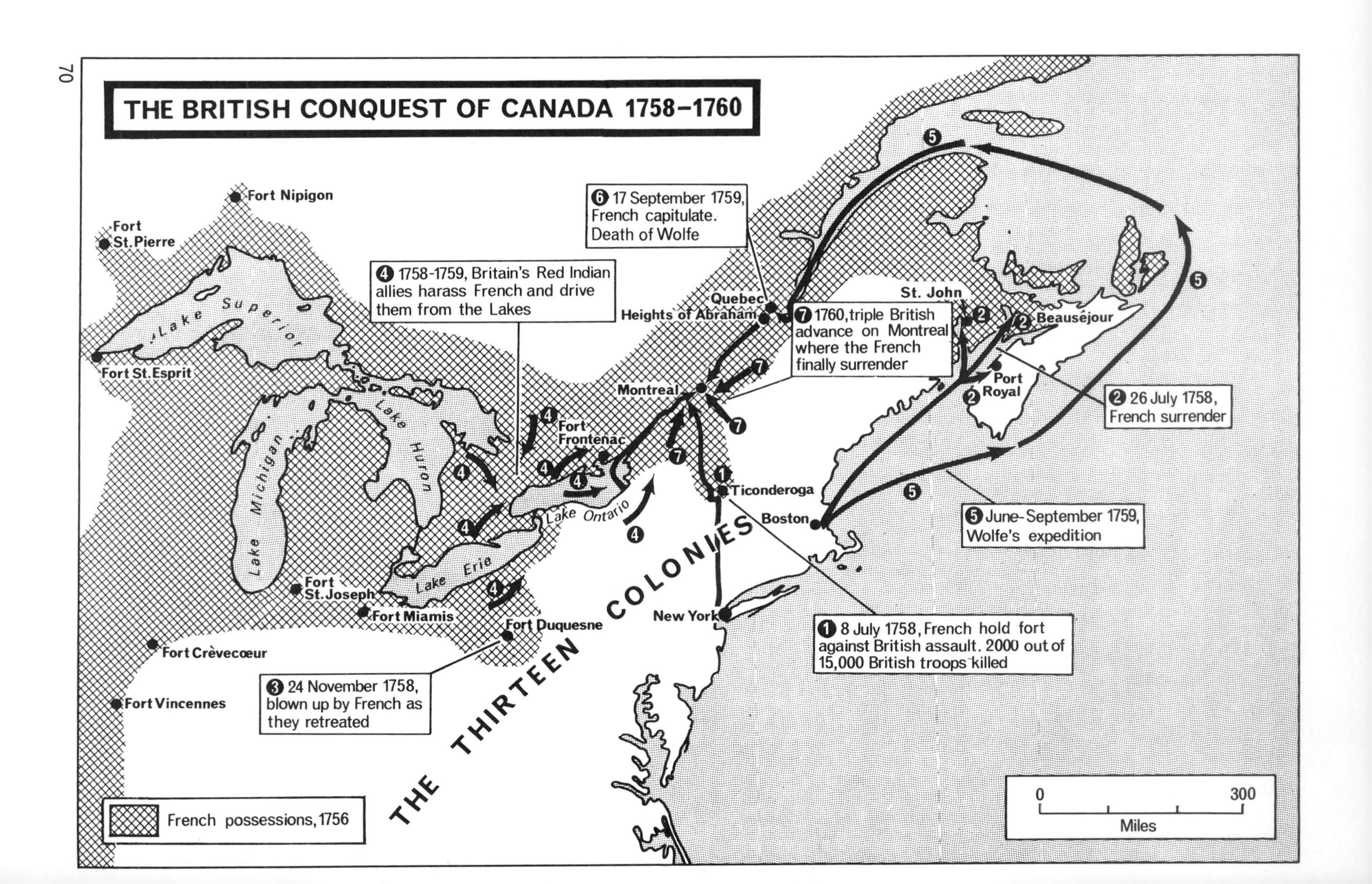
THE BRITISH CONQUEST OF CANADA 1758–1760
Fort Nipigon
Fort St. Pierre
Lake Superior
Fort St. Esprit
Lake Michigan
Lake Huron
Lake Erie
Lake Ontario
Fort St. Joseph
Fort Miamis
Fort Crèvecœur
Fort Vincennes
Fort Duquesne
Fort Frontenac
Montreal
Heights of Abraham
Quebec
Ticonderoga
Boston
New York
St. John
Beauséjour
Port Royal
THE THIRTEEN COLONIES
❹ 1758-1759, Britain's Red Indian allies harass French and drive them from the Lakes
❻ 17 September 1759, French capitulate. Death of Wolfe
❼ 1760, triple British advance on Montreal where the French finally surrender
❷ 26 July 1758, French surrender
❺ June-September 1759, Wolfe's expedition
❶ 8 July 1758, French hold fort against British assault. 2000 out of 15,000 British troops killed
❸ 24 November 1758, blown up by French as they retreated
French possessions, 1756
0
300
Miles

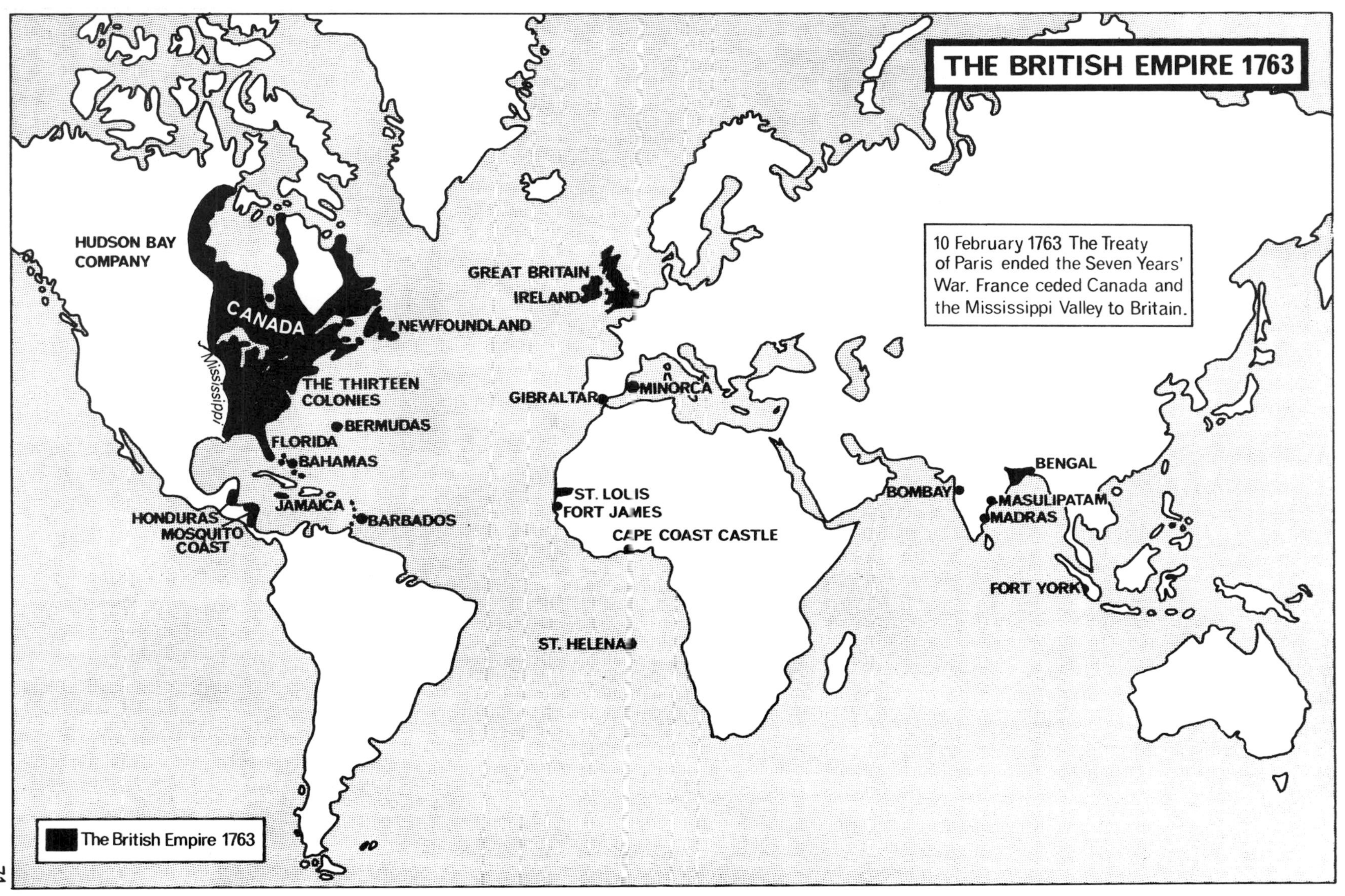
THE BRITISH EMPIRE 1763
10 February 1763 The Treaty of Paris ended the Seven Years' War. France ceded Canada and the Mississippi Valley to Britain.
HUDSON BAY COMPANY
CANADA
Mississippi
NEWFOUNDLAND
THE THIRTEEN COLONIES
BERMUDAS
FLORIDA
BAHAMAS
JAMAICA
BARBADOS
HONDURAS
MOSQUITO COAST
GREAT BRITAIN
IRELAND
GIBRALTAR
MINORCA
ST. LOUIS
FORT JAMES
CAPE COAST CASTLE
ST. HELENA
BOMBAY
BENGAL
MASULIPATAM
MADRAS
FORT YORK
The British Empire 1763

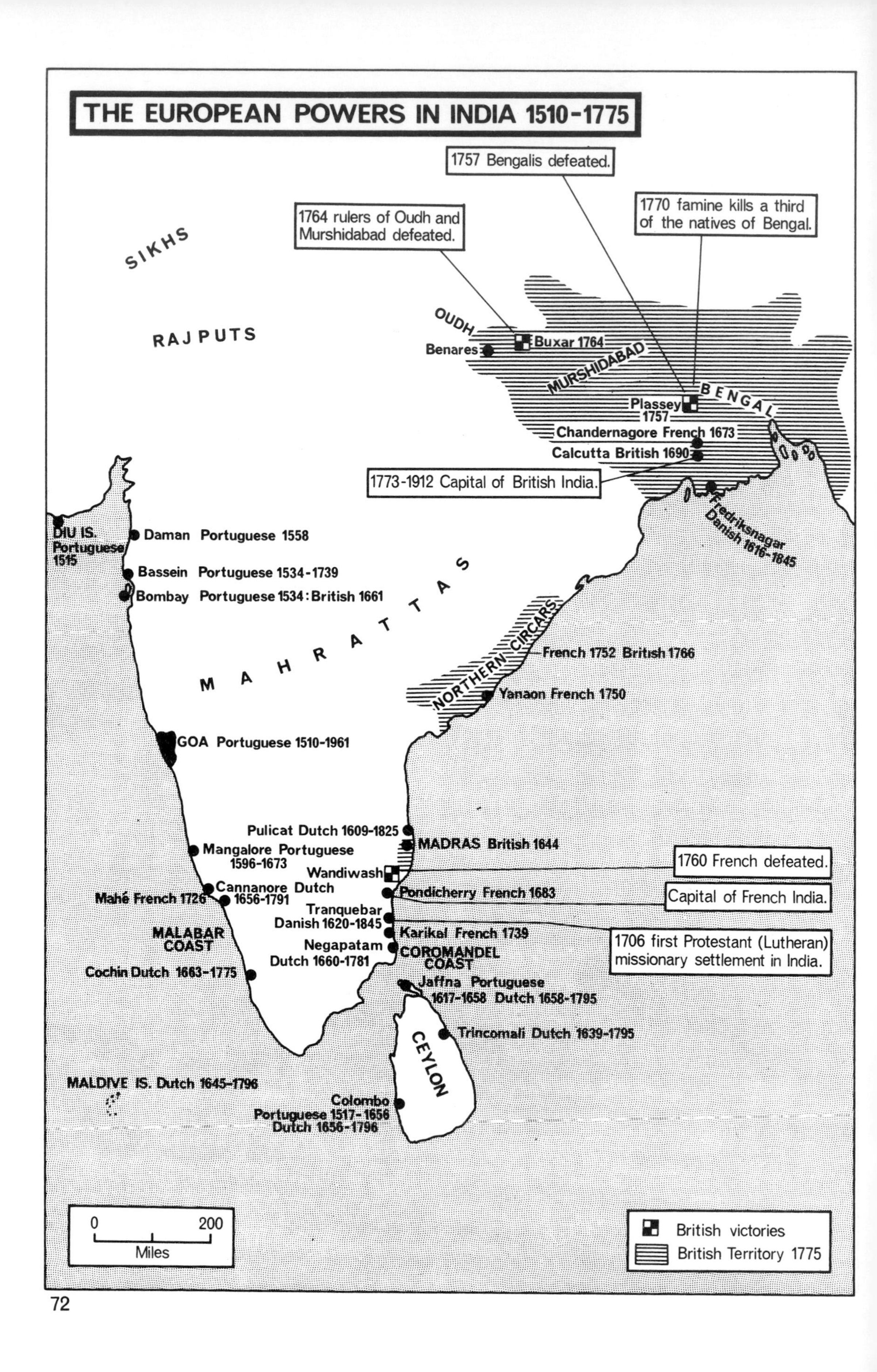
THE EUROPEAN POWERS IN INDIA 1510-1775
1757 Bengalis defeated.
1764 rulers of Oudh and Murshidabad defeated.
1770 famine kills a third of the natives of Bengal.
SIKHS
RAJPUTS
OUDH
Buxar 1764
Benares
MURSHIDABAD
BENGAL
Plassey 1757
Chandernagore French 1673
Calcutta British 1690
1773-1912 Capital of British India.
Fredriksnagar Danish 1616-1845
DIU IS. Portuguese 1515
Daman Portuguese 1558
Bassein Portuguese 1534-1739
Bombay Portuguese 1534: British 1661
MAHRATTAS
NORTHERN CIRCARS
French 1752 British 1766
Yanaon French 1750
GOA Portuguese 1510-1961
Pulicat Dutch 1609-1825
MADRAS British 1644
Mangalore Portuguese 1596-1673
Wandiwash
1760 French defeated.
Cannanore Dutch 1656-1791
Mahé French 1726
Pondicherry French 1683
Capital of French India.
Tranquebar Danish 1620-1845
Karikal French 1739
1706 first Protestant (Lutheran) missionary settlement in India.
MALABAR COAST
Negapatam Dutch 1660-1781
COROMANDEL COAST
Cochin Dutch 1663-1775
Jaffna Portuguese 1617-1658 Dutch 1658-1795
Trincomali Dutch 1639-1795
CEYLON
MALDIVE IS. Dutch 1645-1796
Colombo Portuguese 1517-1656 Dutch 1656-1796
0 200
Miles
British victories
British Territory 1775

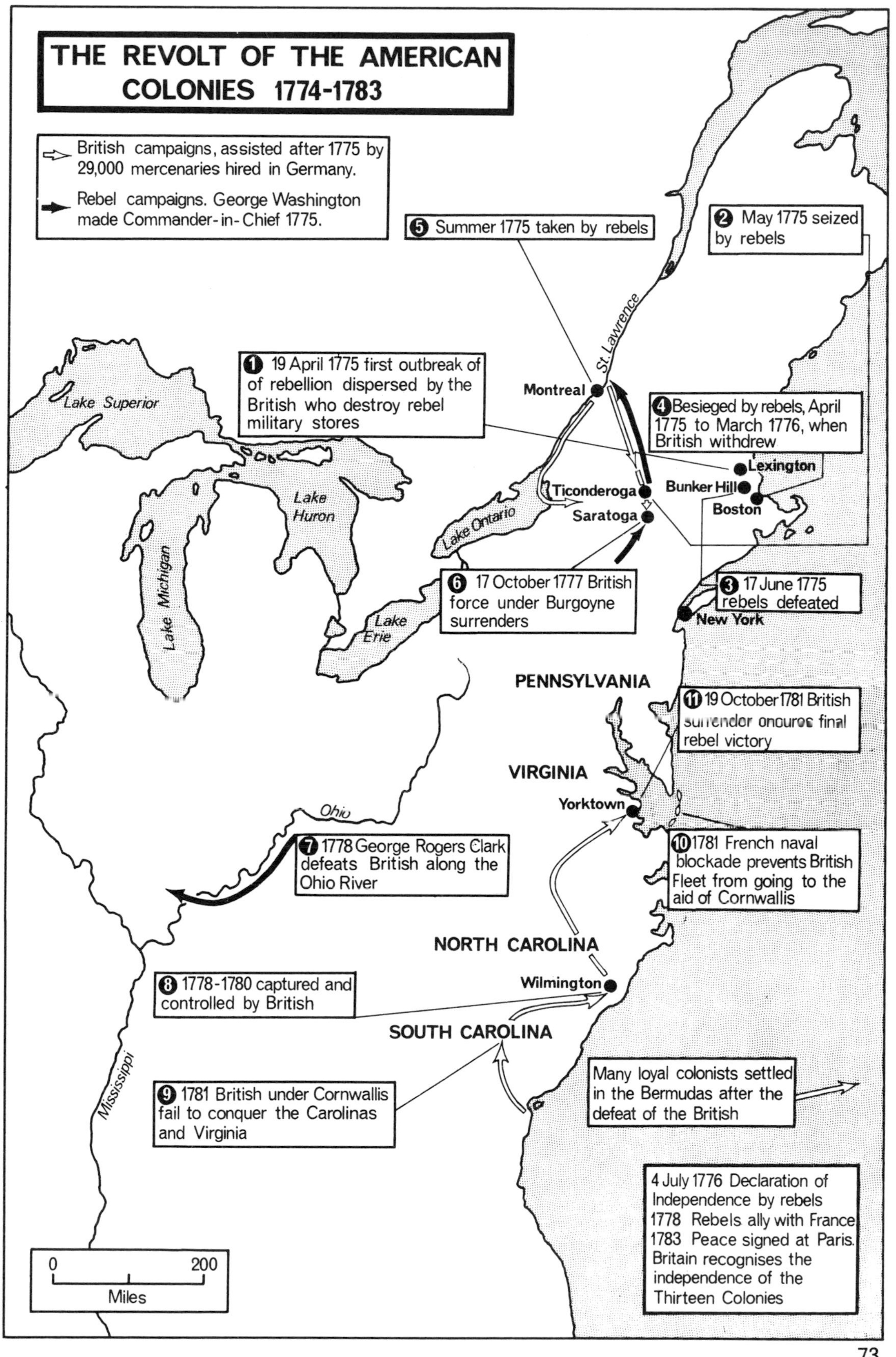
THE REVOLT OF THE AMERICAN COLONIES 1774-1783
British campaigns, assisted after 1775 by 29,000 mercenaries hired in Germany.
Rebel campaigns. George Washington made Commander-in-Chief 1775.
5 Summer 1775 taken by rebels
2 May 1775 seized by rebels
1 19 April 1775 first outbreak of of rebellion dispersed by the British who destroy rebel military stores
4 Besieged by rebels, April 1775 to March 1776, when British withdrew
St. Lawrence
Montreal
Lake Superior
Lake Huron
Lake Michigan
Lake Ontario
Lake Erie
Ticonderoga
Saratoga
Lexington
Bunker Hill
Boston
6 17 October 1777 British force under Burgoyne surrenders
3 17 June 1775 rebels defeated
New York
PENNSYLVANIA
11 19 October 1781 British surrender ensures final rebel victory
VIRGINIA
Yorktown
Ohio
7 1778 George Rogers Clark defeats British along the Ohio River
10 1781 French naval blockade prevents British Fleet from going to the aid of Cornwallis
NORTH CAROLINA
Wilmington
8 1778-1780 captured and controlled by British
SOUTH CAROLINA
Mississippi
9 1781 British under Cornwallis fail to conquer the Carolinas and Virginia
Many loyal colonists settled in the Bermudas after the defeat of the British
4 July 1776 Declaration of Independence by rebels
1778 Rebels ally with France
1783 Peace signed at Paris. Britain recognises the independence of the Thirteen Colonies
0 200
Miles

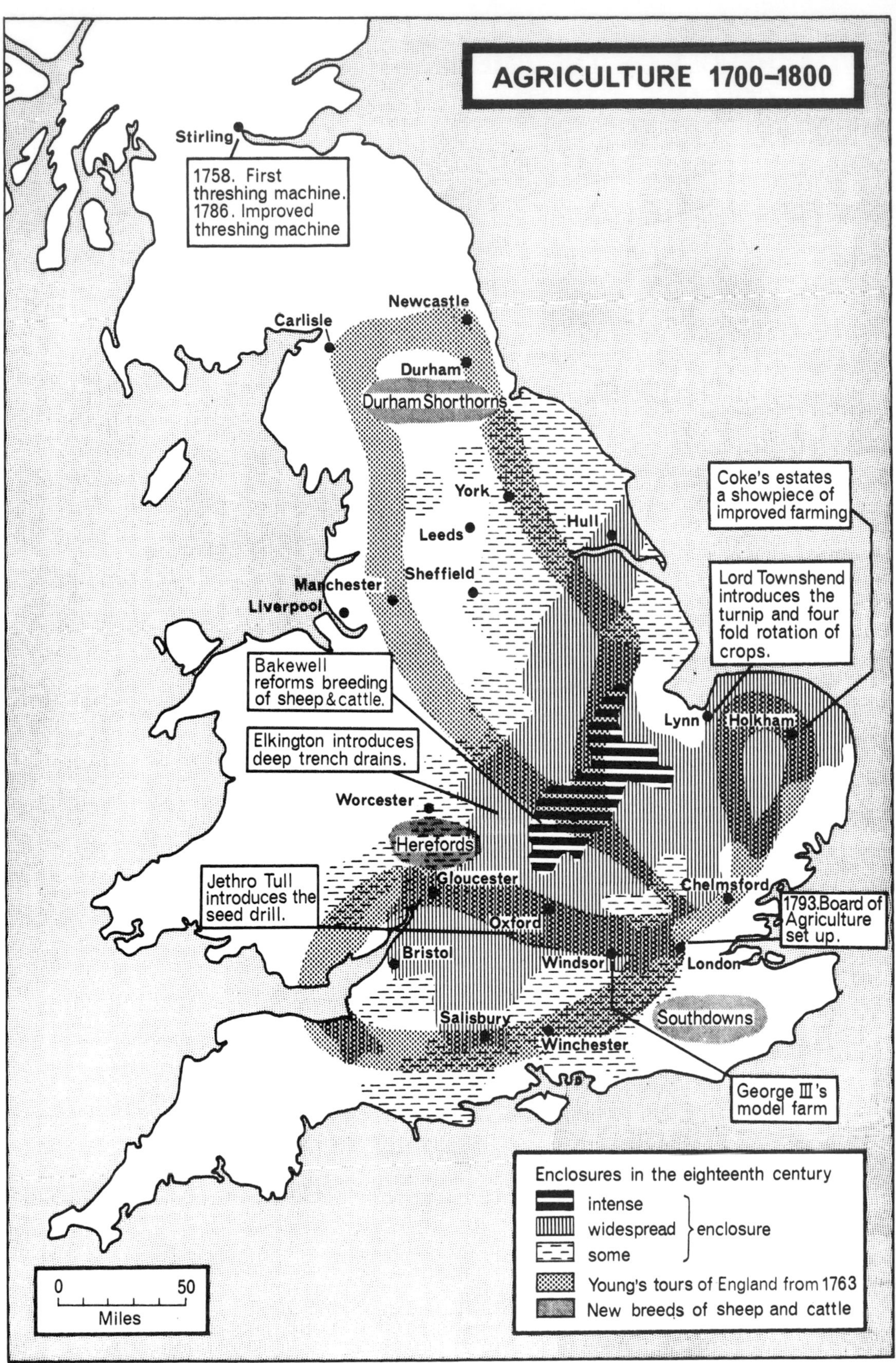

AGRICULTURE 1700–1800
Stirling
1758. First threshing machine. 1786. Improved threshing machine
Newcastle
Carlisle
Durham
Durham Shorthorns
York
Leeds
Hull
Sheffield
Manchester
Liverpool
Coke's estates a showpiece of improved farming
Lord Townshend introduces the turnip and four fold rotation of crops.
Bakewell reforms breeding of sheep & cattle.
Elkington introduces deep trench drains.
Lynn
Holkham
Worcester
Herefords
Jethro Tull introduces the seed drill.
Gloucester
Chelmsford
1793. Board of Agriculture set up.
Oxford
Bristol
Windsor
London
Southdowns
Salisbury
Winchester
George III's model farm
Enclosures in the eighteenth century
intense
widespread
some
enclosure
Young's tours of England from 1763
New breeds of sheep and cattle
0
50
Miles

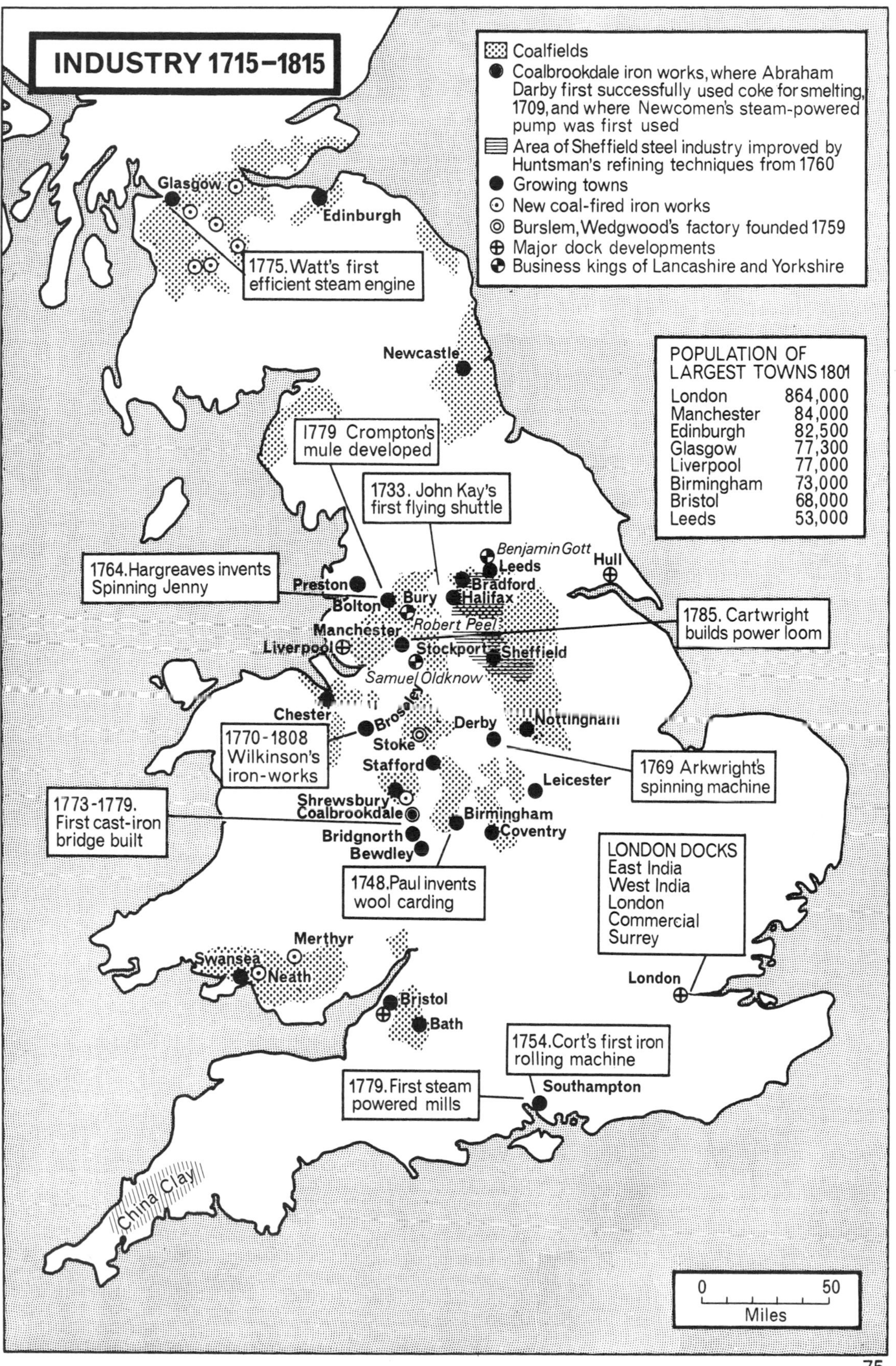
INDUSTRY 1715–1815
Coalfields
Coalbrookdale iron works, where Abraham Darby first successfully used coke for smelting, 1709, and where Newcomen's steam-powered pump was first used
Area of Sheffield steel industry improved by Huntsman's refining techniques from 1760
Growing towns
New coal-fired iron works
Burslem, Wedgwood's factory founded 1759
Major dock developments
Business kings of Lancashire and Yorkshire
POPULATION OF LARGEST TOWNS 1801
London 864,000
Manchester 84,000
Edinburgh 82,500
Glasgow 77,300
Liverpool 77,000
Birmingham 73,000
Bristol 68,000
Leeds 53,000
Glasgow
Edinburgh
1775. Watt's first efficient steam engine
Newcastle
1779 Crompton's mule developed
1733. John Kay's first flying shuttle
1764. Hargreaves invents Spinning Jenny
Benjamin Gott
Leeds
Hull
Bradford
Halifax
Preston
Bury
Bolton
Robert Peel
Manchester
Liverpool
Stockport
Sheffield
1785. Cartwright builds power loom
Samuel Oldknow
Chester
Broseley
Derby
Nottingham
Stoke
1770-1808 Wilkinson's iron-works
Stafford
1769 Arkwright's spinning machine
Leicester
1773-1779. First cast-iron bridge built
Shrewsbury
Coalbrookdale
Birmingham
Coventry
Bridgnorth
Bewdley
1748. Paul invents wool carding
LONDON DOCKS
East India
West India
London
Commercial
Surrey
Merthyr
Swansea
Neath
London
Bristol
Bath
1754. Cort's first iron rolling machine
1779. First steam powered mills
Southampton
China Clay
0 50
Miles

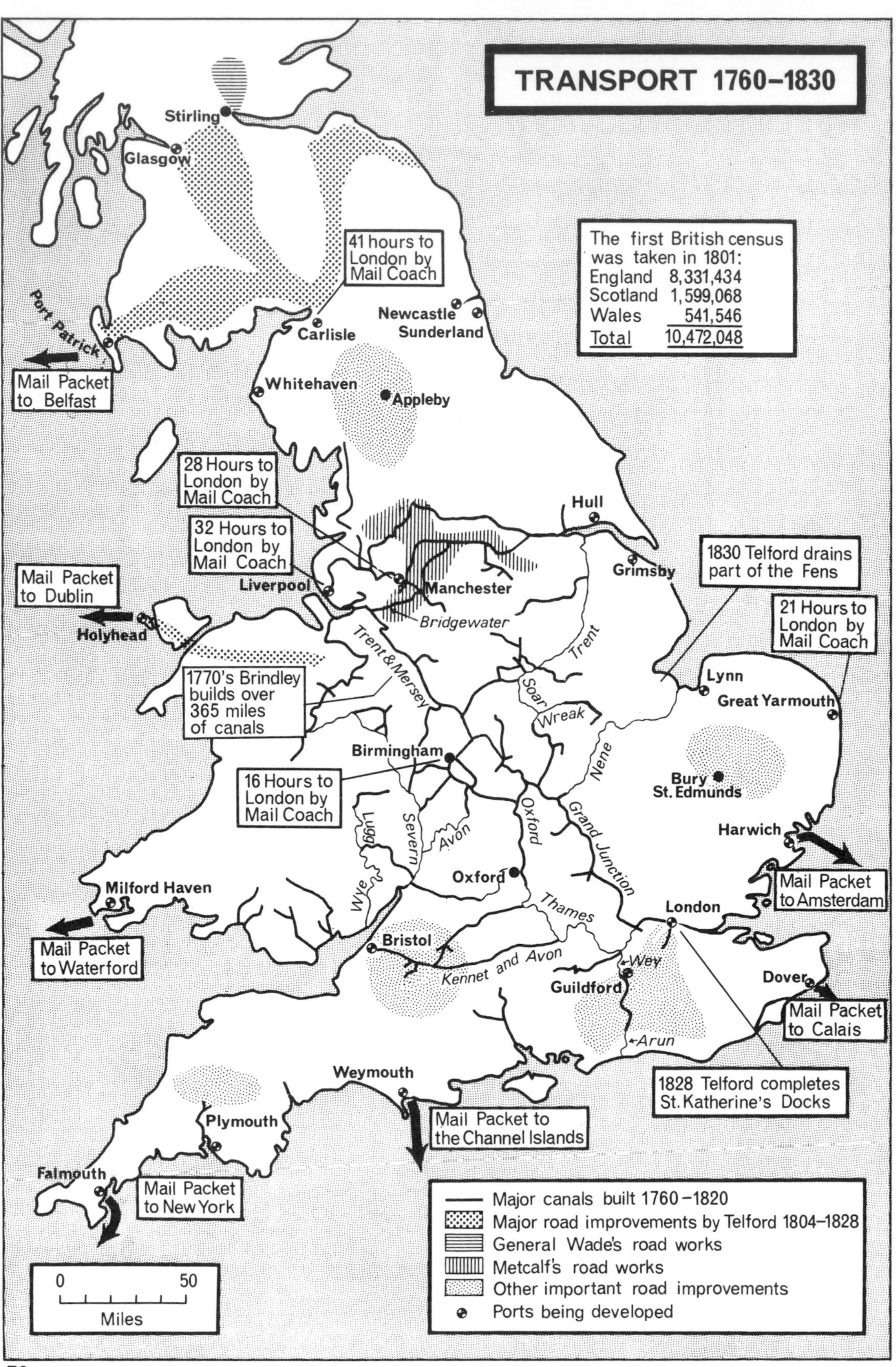
TRANSPORT 1760–1830
Stirling
Glasgow
41 hours to London by Mail Coach
The first British census was taken in 1801:
England 8,331,434
Scotland 1,599,068
Wales 541,546
Total 10,472,048
Port Patrick
Carlisle
Newcastle
Sunderland
Mail Packet to Belfast
Whitehaven
Appleby
28 Hours to London by Mail Coach
Hull
32 Hours to London by Mail Coach
Grimsby
1830 Telford drains part of the Fens
Mail Packet to Dublin
Liverpool
Manchester
Holyhead
Bridgewater
Trent
21 Hours to London by Mail Coach
Trent & Mersey
Lynn
1770's Brindley builds over 365 miles of canals
Soar
Wreak
Great Yarmouth
Birmingham
Nene
16 Hours to London by Mail Coach
Bury St. Edmunds
Lugg
Severn
Avon
Oxford
Grand Junction
Harwich
Wye
Oxford
Mail Packet to Amsterdam
Milford Haven
Thames
London
Mail Packet to Waterford
Bristol
Kennet and Avon
Wey
Guildford
Dover
Mail Packet to Calais
Arun
1828 Telford completes St. Katherine's Docks
Weymouth
Mail Packet to the Channel Islands
Plymouth
Falmouth
Mail Packet to New York
Major canals built 1760–1820
Major road improvements by Telford 1804–1828
General Wade's road works
Metcalf's road works
Other important road improvements
Ports being developed
0
50
Miles

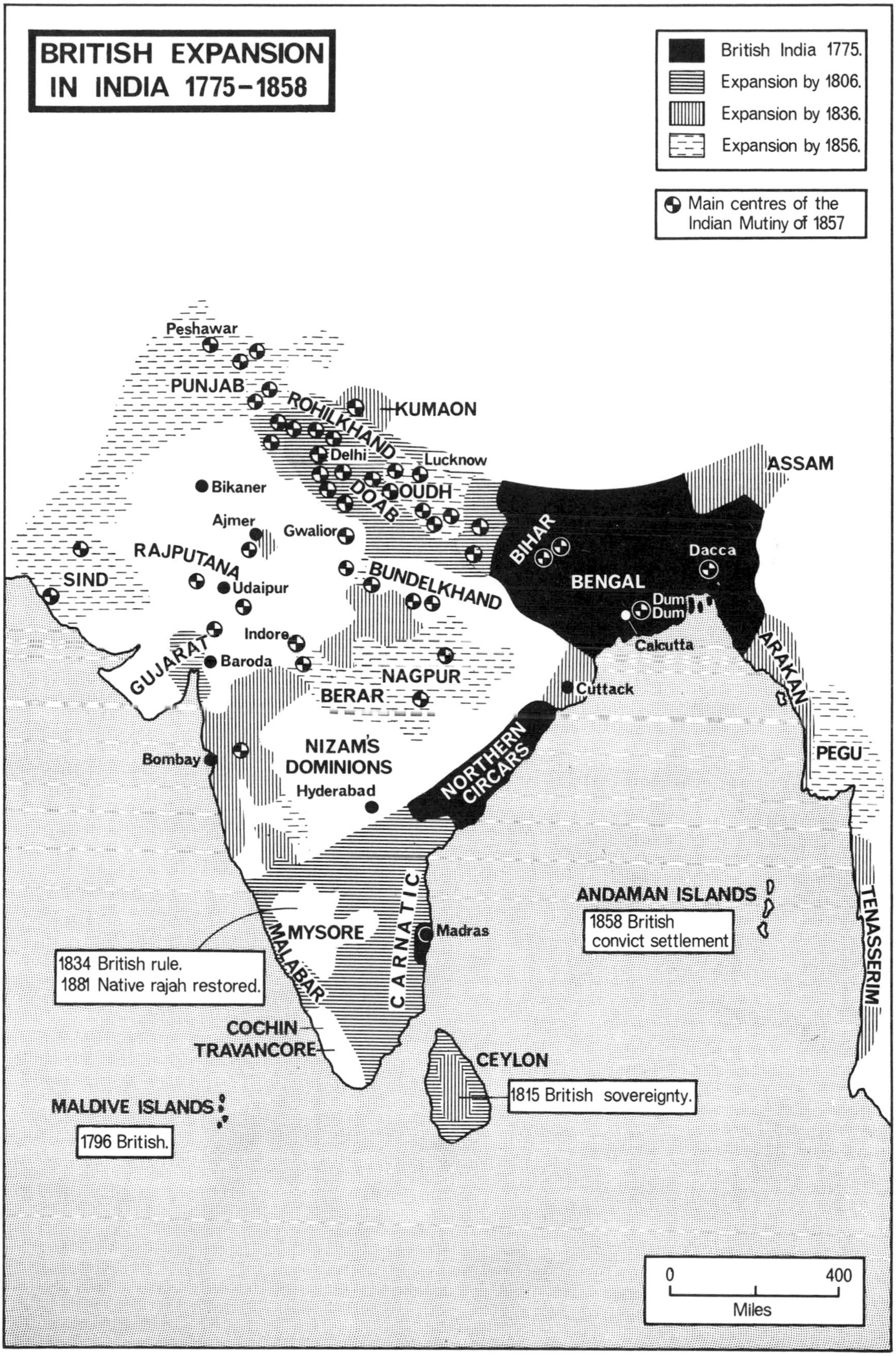

BRITISH EXPANSION IN INDIA 1775–1858

British India 1775.
Expansion by 1806.
Expansion by 1836.
Expansion by 1856.

Main centres of the Indian Mutiny of 1857

Peshawar
PUNJAB
ROHILKHAND
KUMAON
Delhi
Lucknow
ASSAM
Bikaner
DOAB
OUDH
Ajmer
Gwalior
RAJPUTANA
BIHAR
Dacca
SIND
BUNDELKHAND
BENGAL
Udaipur
Dum Dum
Indore
Calcutta
GUJARAT
Baroda
ARAKAN
NAGPUR
BERAR
Cuttack
NIZAM'S DOMINIONS
Bombay
Hyderabad
NORTHERN CIRCARS
PEGU
ANDAMAN ISLANDS
1858 British convict settlement
CARNATIC
MYSORE
Madras
MALABAR
1834 British rule.
1881 Native rajah restored.
TENASSERIM
COCHIN
TRAVANCORE
CEYLON
1815 British sovereignty.
MALDIVE ISLANDS
1796 British.

0 400
Miles

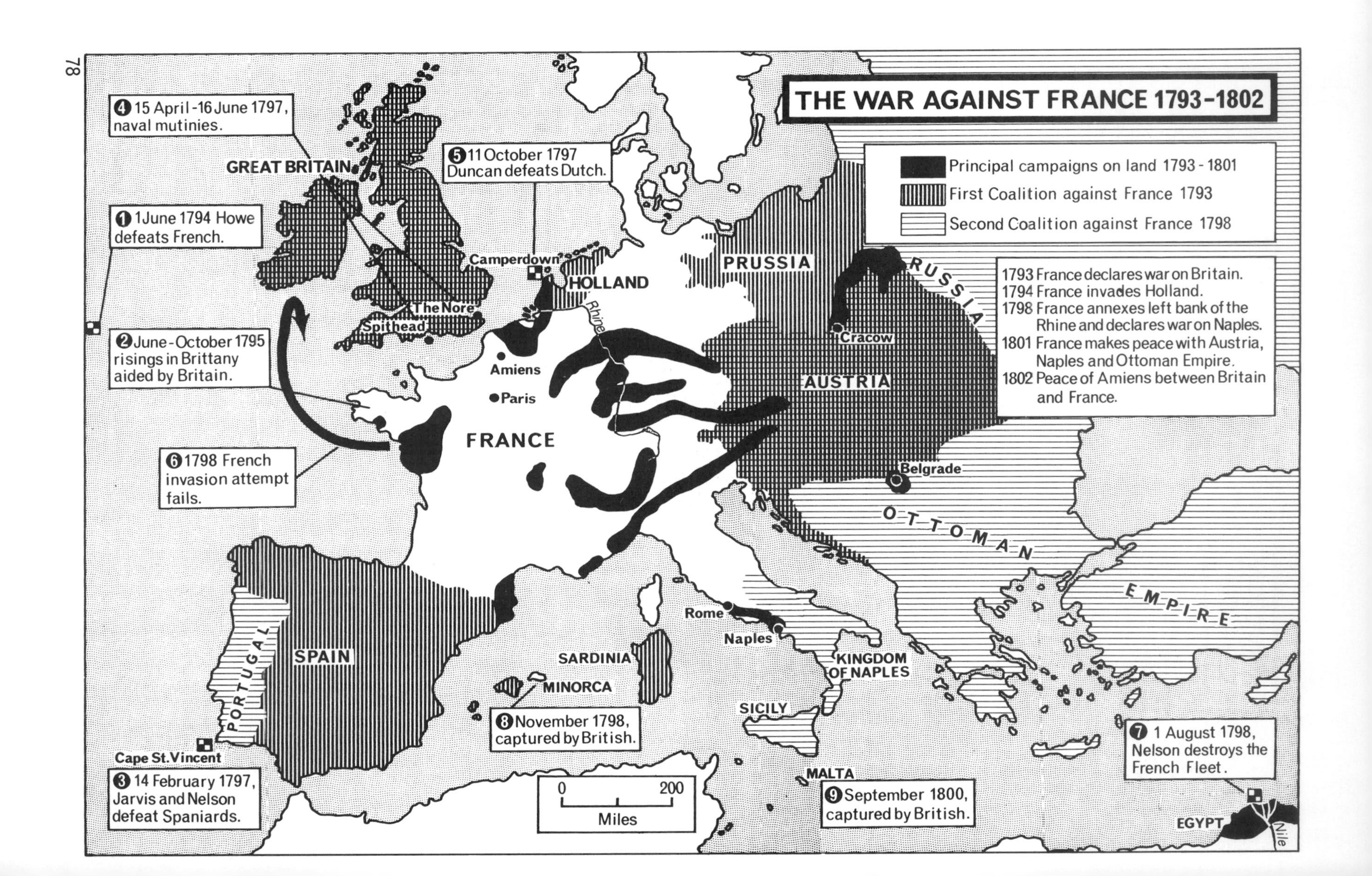
THE WAR AGAINST FRANCE 1793-1802
Principal campaigns on land 1793 - 1801
First Coalition against France 1793
Second Coalition against France 1798
1793 France declares war on Britain.
1794 France invades Holland.
1798 France annexes left bank of the Rhine and declares war on Naples.
1801 France makes peace with Austria, Naples and Ottoman Empire.
1802 Peace of Amiens between Britain and France.
4 15 April - 16 June 1797, naval mutinies.
1 1 June 1794 Howe defeats French.
2 June - October 1795 risings in Brittany aided by Britain.
6 1798 French invasion attempt fails.
5 11 October 1797 Duncan defeats Dutch.
3 14 February 1797, Jarvis and Nelson defeat Spaniards.
8 November 1798, captured by British.
9 September 1800, captured by British.
7 1 August 1798, Nelson destroys the French Fleet.
GREAT BRITAIN
The Nore
Spithead
Camperdown
HOLLAND
Rhine
PRUSSIA
RUSSIA
Cracow
AUSTRIA
Belgrade
OTTOMAN
EMPIRE
Amiens
Paris
FRANCE
SPAIN
PORTUGAL
Cape St. Vincent
SARDINIA
MINORCA
Rome
Naples
KINGDOM OF NAPLES
SICILY
MALTA
EGYPT
Nile
0
200
Miles

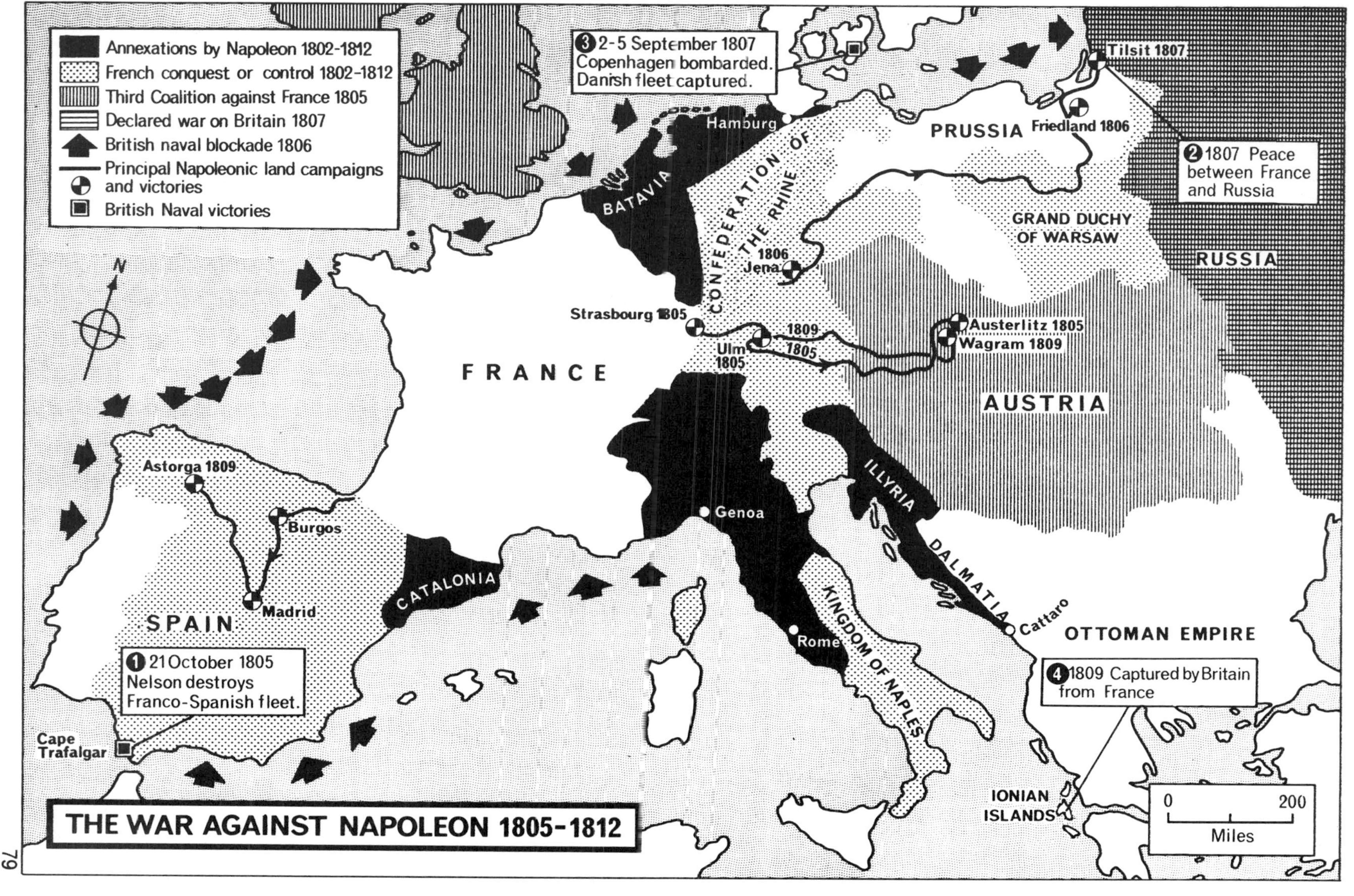

THE WAR AGAINST NAPOLEON 1805-1812

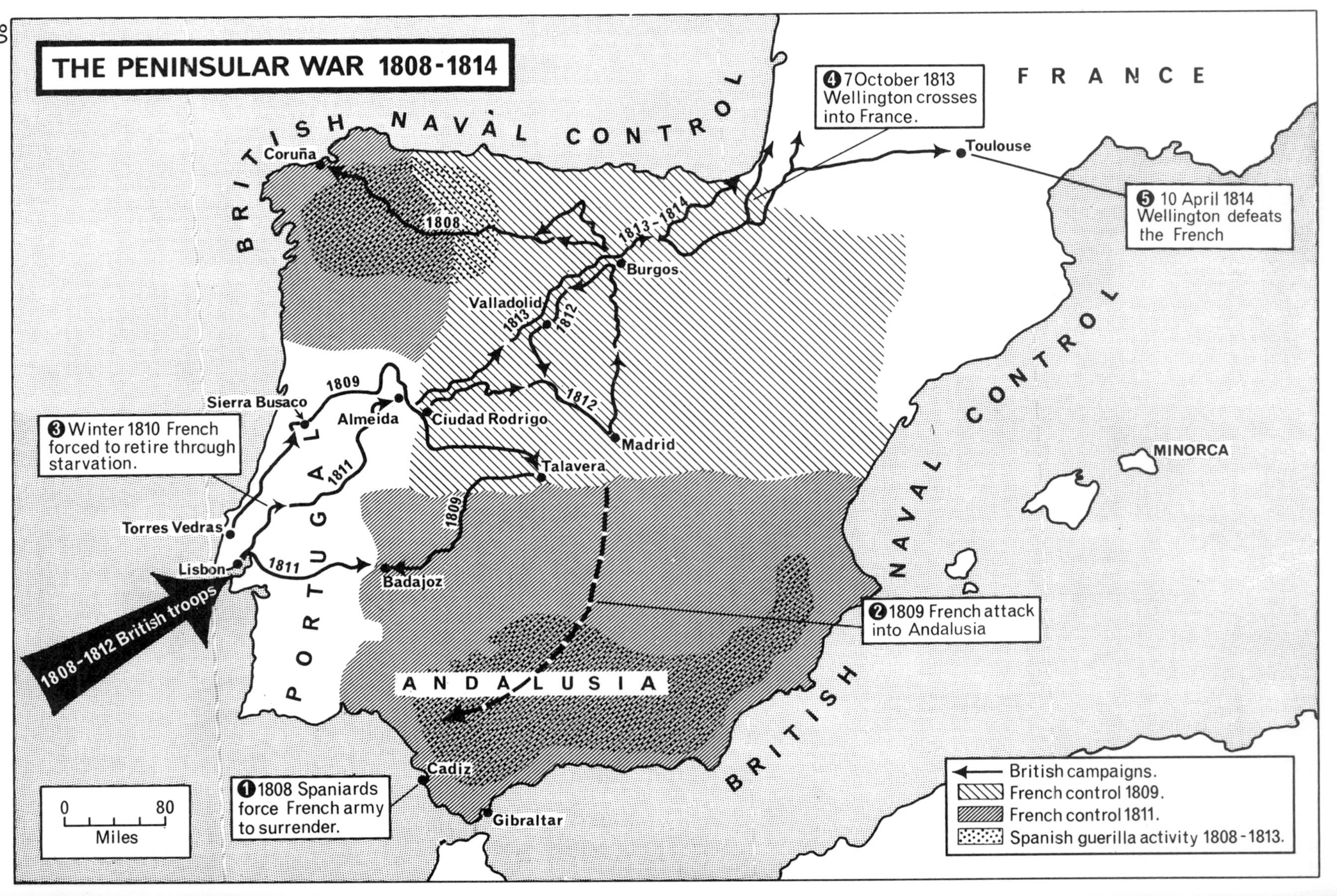

THE PENINSULAR WAR 1808-1814
BRITISH NAVAL CONTROL
FRANCE
❹ 7 October 1813 Wellington crosses into France.
❺ 10 April 1814 Wellington defeats the French
Toulouse
Coruña
1808
1813-1814
Burgos
Valladolid
1813
1812
1812
Madrid
Sierra Busaco
1809
Almeida
Ciudad Rodrigo
❸ Winter 1810 French forced to retire through starvation.
1811
Talavera
MINORCA
Torres Vedras
Lisbon
1811
1809
Badajoz
PORTUGAL
1808-1812 British troops
❷ 1809 French attack into Andalusia
ANDALUSIA
Cadiz
Gibraltar
❶ 1808 Spaniards force French army to surrender.
0
80
Miles
British campaigns.
French control 1809.
French control 1811.
Spanish guerilla activity 1808-1813.

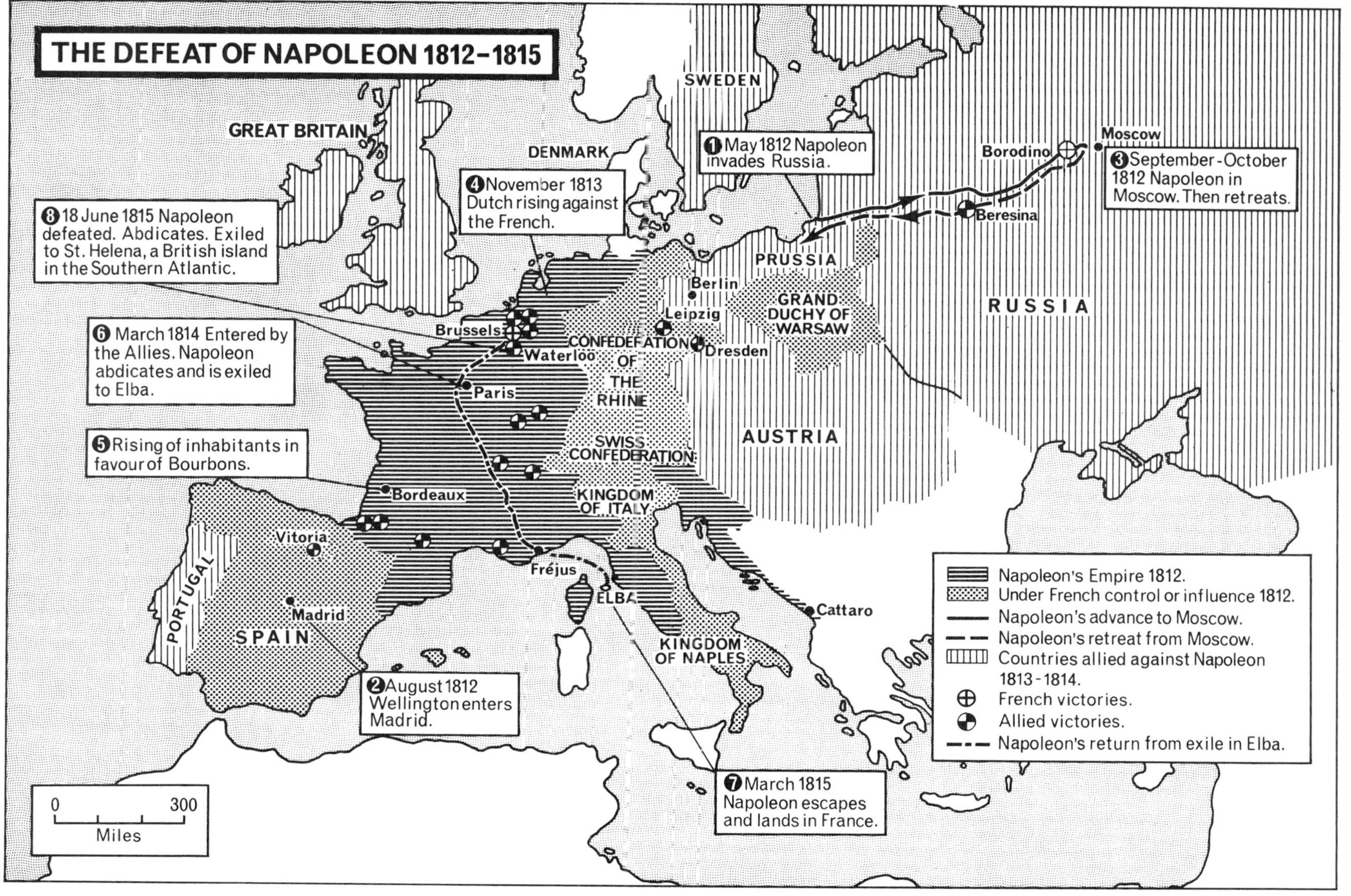
THE DEFEAT OF NAPOLEON 1812–1815
GREAT BRITAIN
DENMARK
SWEDEN
PRUSSIA
RUSSIA
GRAND DUCHY OF WARSAW
AUSTRIA
CONFEDERATION OF THE RHINE
SWISS CONFEDERATION
KINGDOM OF ITALY
KINGDOM OF NAPLES
SPAIN
PORTUGAL
ELBA
Moscow
Borodino
Beresina
Berlin
Leipzig
Dresden
Brussels
Waterloo
Paris
Bordeaux
Vitoria
Madrid
Fréjus
Cattaro
1 May 1812 Napoleon invades Russia.
2 August 1812 Wellington enters Madrid.
3 September-October 1812 Napoleon in Moscow. Then retreats.
4 November 1813 Dutch rising against the French.
5 Rising of inhabitants in favour of Bourbons.
6 March 1814 Entered by the Allies. Napoleon abdicates and is exiled to Elba.
7 March 1815 Napoleon escapes and lands in France.
8 18 June 1815 Napoleon defeated. Abdicates. Exiled to St. Helena, a British island in the Southern Atlantic.
Napoleon's Empire 1812.
Under French control or influence 1812.
Napoleon's advance to Moscow.
Napoleon's retreat from Moscow.
Countries allied against Napoleon 1813-1814.
French victories.
Allied victories.
Napoleon's return from exile in Elba.
0
300
Miles

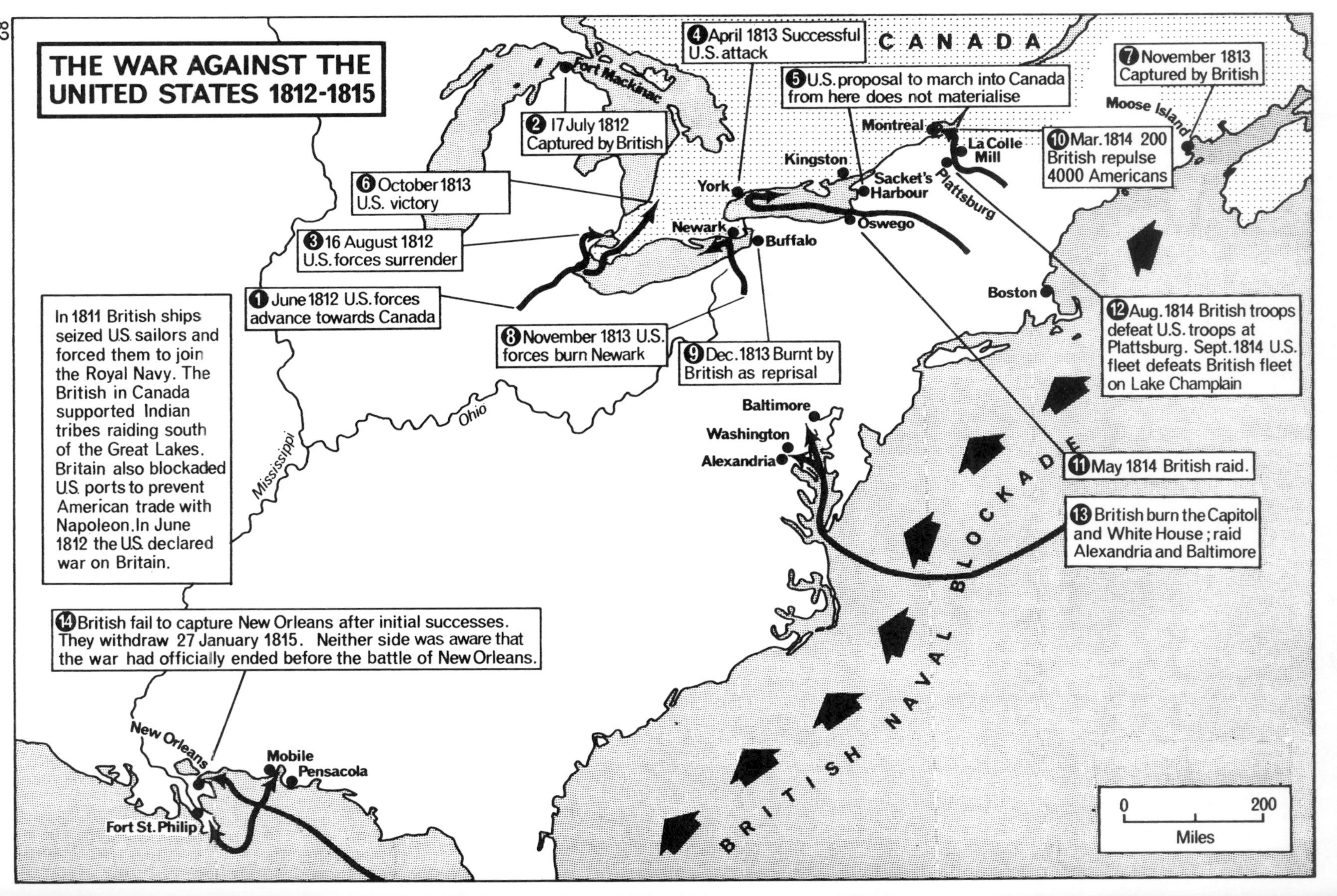
THE WAR AGAINST THE UNITED STATES 1812-1815
CANADA
1 June 1812 U.S. forces advance towards Canada
2 17 July 1812 Captured by British
3 16 August 1812 U.S. forces surrender
4 April 1813 Successful U.S. attack
5 U.S. proposal to march into Canada from here does not materialise
6 October 1813 U.S. victory
7 November 1813 Captured by British
8 November 1813 U.S. forces burn Newark
9 Dec. 1813 Burnt by British as reprisal
10 Mar. 1814 200 British repulse 4000 Americans
11 May 1814 British raid.
12 Aug. 1814 British troops defeat U.S. troops at Plattsburg. Sept. 1814 U.S. fleet defeats British fleet on Lake Champlain
13 British burn the Capitol and White House; raid Alexandria and Baltimore
14 British fail to capture New Orleans after initial successes. They withdraw 27 January 1815. Neither side was aware that the war had officially ended before the battle of New Orleans.
In 1811 British ships seized U.S. sailors and forced them to join the Royal Navy. The British in Canada supported Indian tribes raiding south of the Great Lakes. Britain also blockaded U.S. ports to prevent American trade with Napoleon. In June 1812 the U.S. declared war on Britain.
Fort Mackinac
York
Kingston
Montreal
La Colle Mill
Sacket's Harbour
Plattsburg
Oswego
Newark
Buffalo
Moose Island
Boston
Baltimore
Washington
Alexandria
Ohio
Mississippi
New Orleans
Mobile
Pensacola
Fort St. Philip
BRITISH NAVAL BLOCKADE
0
200
Miles

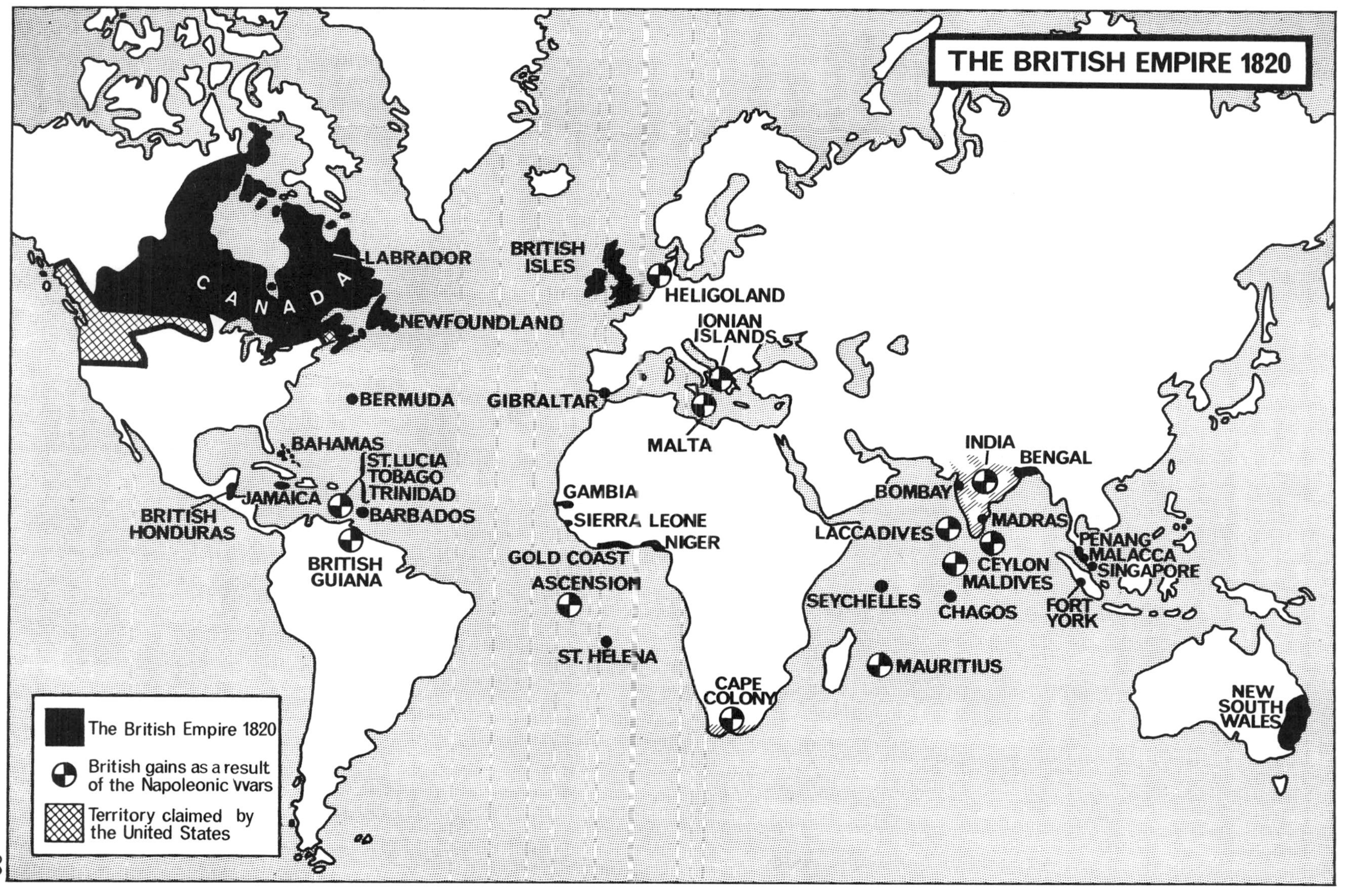
THE BRITISH EMPIRE 1820
CANADA
LABRADOR
NEWFOUNDLAND
BRITISH ISLES
HELIGOLAND
IONIAN ISLANDS
MALTA
GIBRALTAR
BERMUDA
BAHAMAS
ST. LUCIA
TOBAGO
TRINIDAD
BARBADOS
JAMAICA
BRITISH HONDURAS
BRITISH GUIANA
GAMBIA
SIERRA LEONE
NIGER
GOLD COAST
ASCENSION
ST. HELENA
CAPE COLONY
INDIA
BENGAL
BOMBAY
MADRAS
LACCADIVES
CEYLON
MALDIVES
CHAGOS
SEYCHELLES
MAURITIUS
PENANG
MALACCA
SINGAPORE
FORT YORK
NEW SOUTH WALES
The British Empire 1820
British gains as a result of the Napoleonic Wars
Territory claimed by the United States

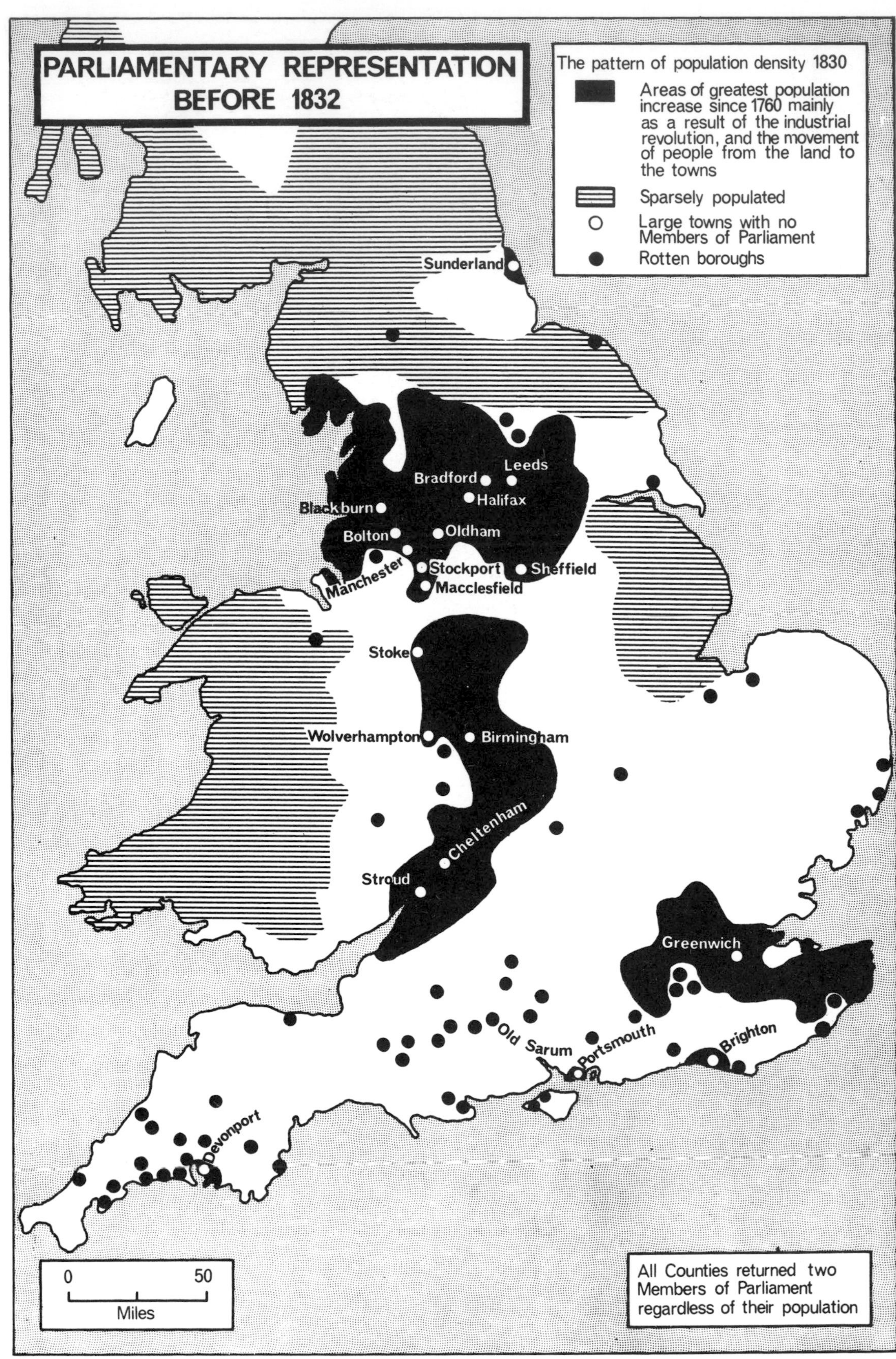

PARLIAMENTARY REPRESENTATION
BEFORE 1832
The pattern of population density 1830
Areas of greatest population increase since 1760 mainly as a result of the industrial revolution, and the movement of people from the land to the towns
Sparsely populated
Large towns with no Members of Parliament
Rotten boroughs
Sunderland
Leeds
Bradford
Halifax
Blackburn
Bolton
Oldham
Manchester
Stockport
Sheffield
Macclesfield
Stoke
Wolverhampton
Birmingham
Cheltenham
Stroud
Greenwich
Old Sarum
Portsmouth
Brighton
Devonport
0
50
Miles
All Counties returned two Members of Parliament regardless of their population

PARLIAMENTARY REFORM 1832
NORTHUMBERLAND
Tynemouth
South Shields
Gateshead
Sunderland
DURHAM
CUMBERLAND
Whitehaven
Kendal
Whitby
YORKSHIRE
LANCASHIRE
Blackburn
Bolton
Bury
Oldham
Leeds
Bradford
Halifax
Wakefield
Huddersfield
Salford
Ashton
Stockport
Manchester
Sheffield
Warrington
Macclesfield
CHESHIRE
LINCOLN
NOTTINGHAM
Stoke
DERBY
STAFFORD
SHROPSHIRE
LEICESTER
NORFOLK
Walsall
Birmingham
Wolverhampton
Dudley
Kidderminster
WORCS
WARWICK
NORTHAMPTON
CAMBRIDGE
SUFFOLK
HEREFORD
Cheltenham
Stroud
GLOUCESTER
OXFORD
BUCKINGHAM
HERTFORD
ESSEX
Marylebone
Tower Hamlets
Finsbury
Greenwich
Chatham
Lambeth
BERKSHIRE
Merthyr Tydfil
WILTSHIRE
SURREY
KENT
Frome
HAMPSHIRE
SUSSEX
SOMERSET
Brighton
DORSET
ISLE OF WIGHT
DEVON
Devonport
CORNWALL
Towns enfranchised with two Members of Parliament
Towns enfranchised with one Member of Parliament
Counties gaining two extra Members of Parliament
Counties gaining one extra Member of Parliament
0
50
Miles

F

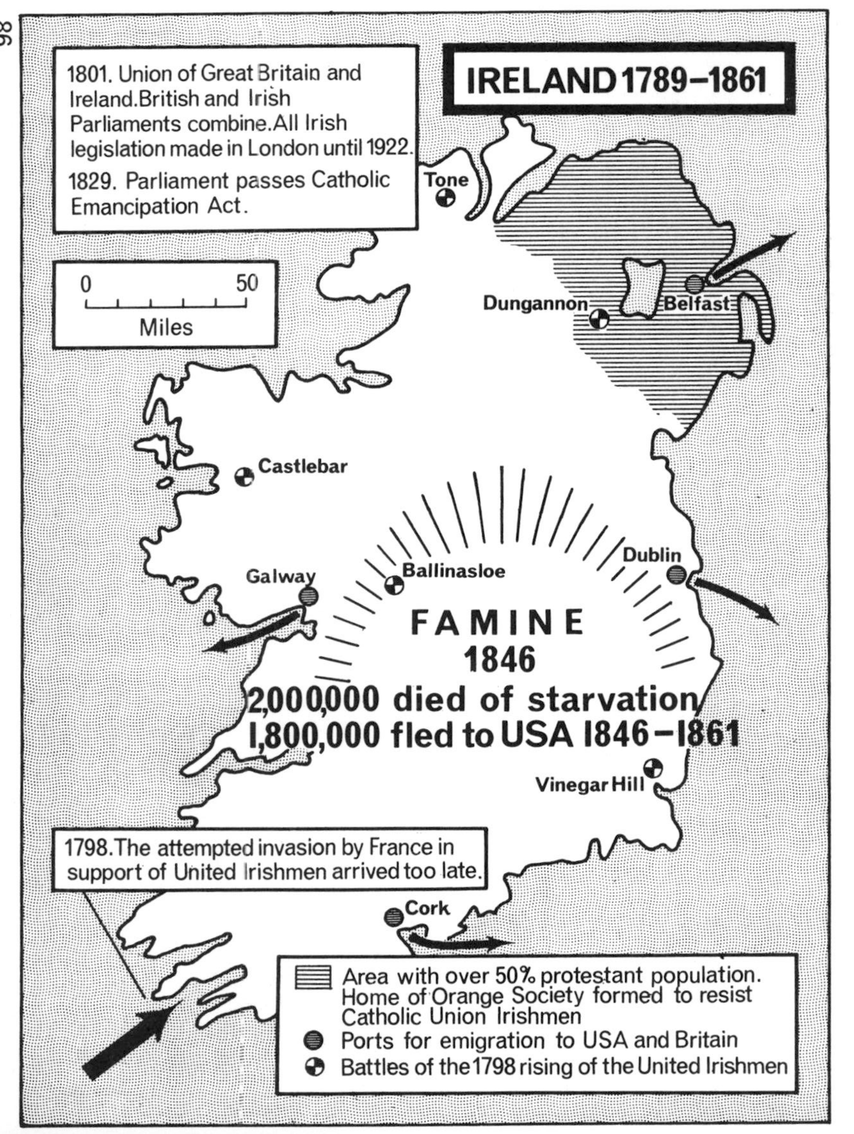

IRELAND 1789–1861
1801. Union of Great Britain and Ireland. British and Irish Parliaments combine. All Irish legislation made in London until 1922.
1829. Parliament passes Catholic Emancipation Act.
0 50
Miles
Tone
Dungannon
Belfast
Castlebar
Galway
Ballinasloe
Dublin
FAMINE
1846
2,000,000 died of starvation
1,800,000 fled to USA 1846–1861
Vinegar Hill
1798. The attempted invasion by France in support of United Irishmen arrived too late.
Cork
Area with over 50% protestant population. Home of Orange Society formed to resist Catholic Union Irishmen
Ports for emigration to USA and Britain
Battles of the 1798 rising of the United Irishmen

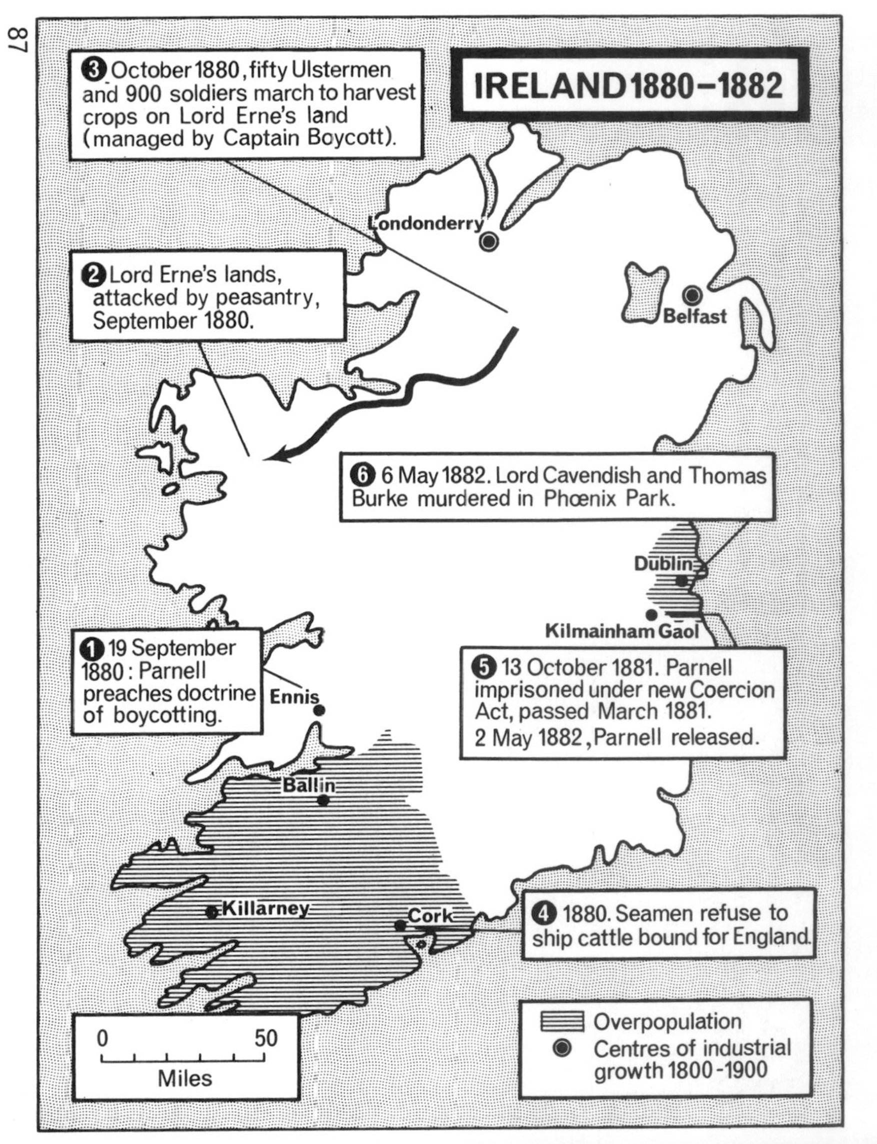

IRELAND 1880–1882
3 October 1880, fifty Ulstermen and 900 soldiers march to harvest crops on Lord Erne's land (managed by Captain Boycott).
Londonderry
Belfast
2 Lord Erne's lands, attacked by peasantry, September 1880.
6 6 May 1882. Lord Cavendish and Thomas Burke murdered in Phœnix Park.
Dublin
Kilmainham Gaol
1 19 September 1880: Parnell preaches doctrine of boycotting.
Ennis
5 13 October 1881. Parnell imprisoned under new Coercion Act, passed March 1881.
2 May 1882, Parnell released.
Ballin
Killarney
Cork
4 1880. Seamen refuse to ship cattle bound for England.
0 50
Miles
Overpopulation
Centres of industrial growth 1800-1900

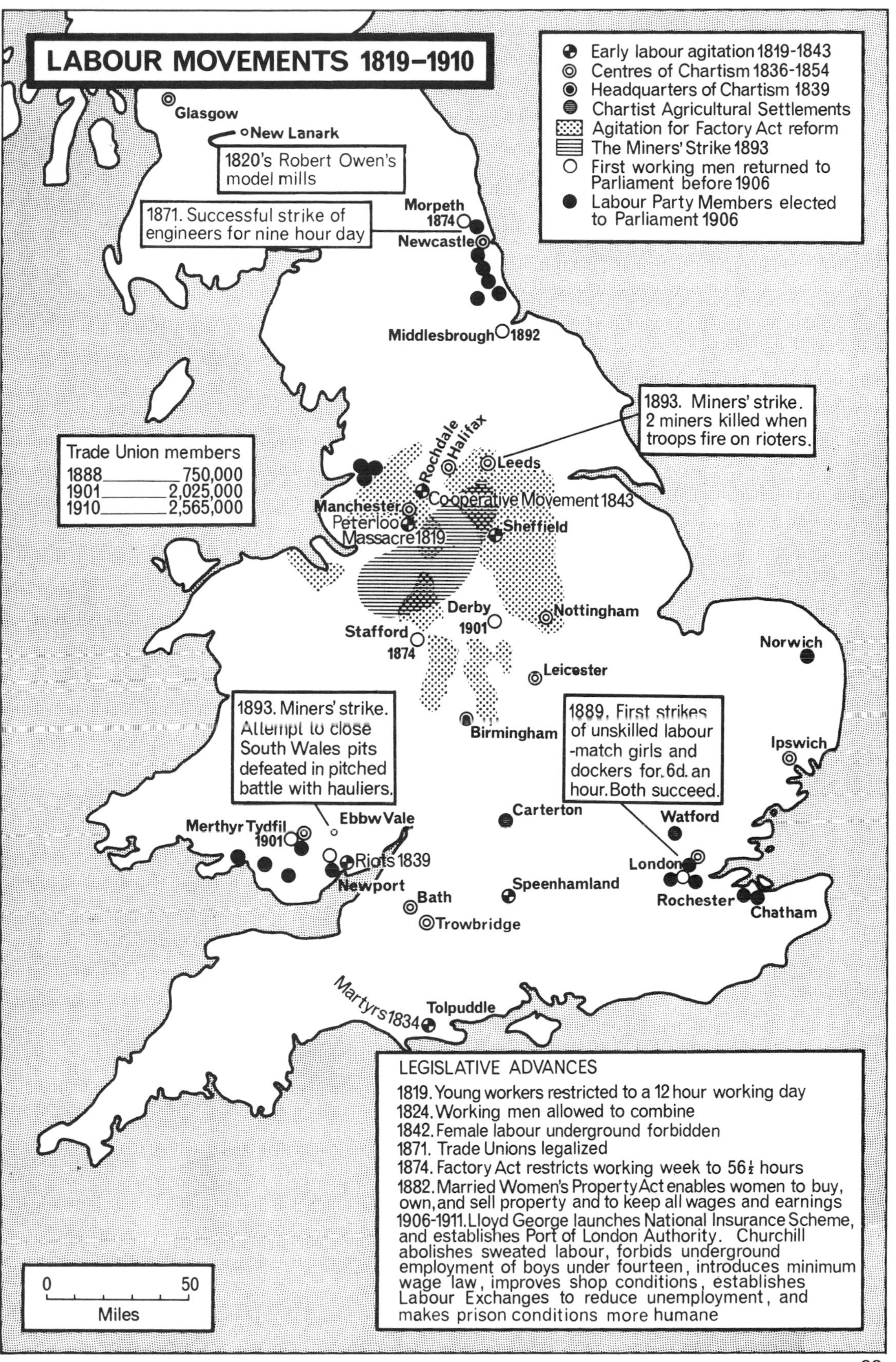

LABOUR MOVEMENTS 1819–1910
Early labour agitation 1819-1843
Centres of Chartism 1836-1854
Headquarters of Chartism 1839
Chartist Agricultural Settlements
Agitation for Factory Act reform
The Miners' Strike 1893
First working men returned to Parliament before 1906
Labour Party Members elected to Parliament 1906
Glasgow
New Lanark
1820's Robert Owen's model mills
1871. Successful strike of engineers for nine hour day
Morpeth
1874
Newcastle
Middlesbrough 1892
1893. Miners' strike. 2 miners killed when troops fire on rioters.
Trade Union members
1888 750,000
1901 2,025,000
1910 2,565,000
Rochdale
Halifax
Leeds
Co-operative Movement 1843
Manchester
Peterloo Massacre 1819
Sheffield
Derby 1901
Nottingham
Stafford 1874
Norwich
Leicester
Birmingham
1893. Miners' strike. Attempt to close South Wales pits defeated in pitched battle with hauliers.
1889. First strikes of unskilled labour -match girls and dockers for 6d. an hour. Both succeed.
Ipswich
Carterton
Watford
Merthyr Tydfil 1901
Ebbw Vale
Riots 1839
Newport
London
Rochester
Chatham
Speenhamland
Bath
Trowbridge
Martyrs 1834
Tolpuddle
LEGISLATIVE ADVANCES
1819. Young workers restricted to a 12 hour working day
1824. Working men allowed to combine
1842. Female labour underground forbidden
1871. Trade Unions legalized
1874. Factory Act restricts working week to 56½ hours
1882. Married Women's Property Act enables women to buy, own, and sell property and to keep all wages and earnings
1906-1911. Lloyd George launches National Insurance Scheme, and establishes Port of London Authority. Churchill abolishes sweated labour, forbids underground employment of boys under fourteen, introduces minimum wage law, improves shop conditions, establishes Labour Exchanges to reduce unemployment, and makes prison conditions more humane
0 50
Miles

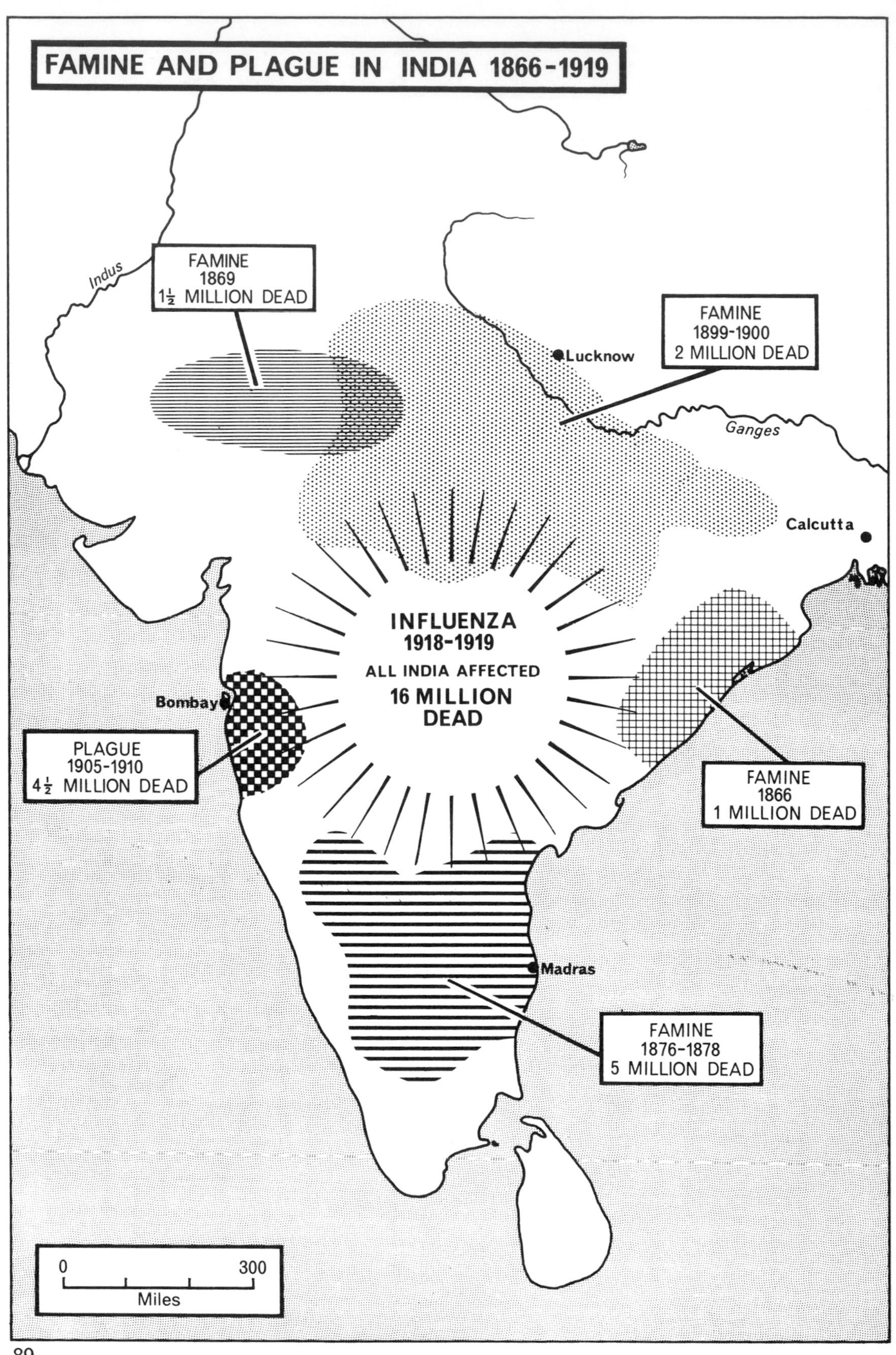
FAMINE AND PLAGUE IN INDIA 1866-1919
Indus
FAMINE
1869
1½ MILLION DEAD
FAMINE
1899-1900
2 MILLION DEAD
Lucknow
Ganges
Calcutta
INFLUENZA
1918-1919
ALL INDIA AFFECTED
16 MILLION
DEAD
Bombay
PLAGUE
1905-1910
4½ MILLION DEAD
FAMINE
1866
1 MILLION DEAD
Madras
FAMINE
1876-1878
5 MILLION DEAD
0
300
Miles

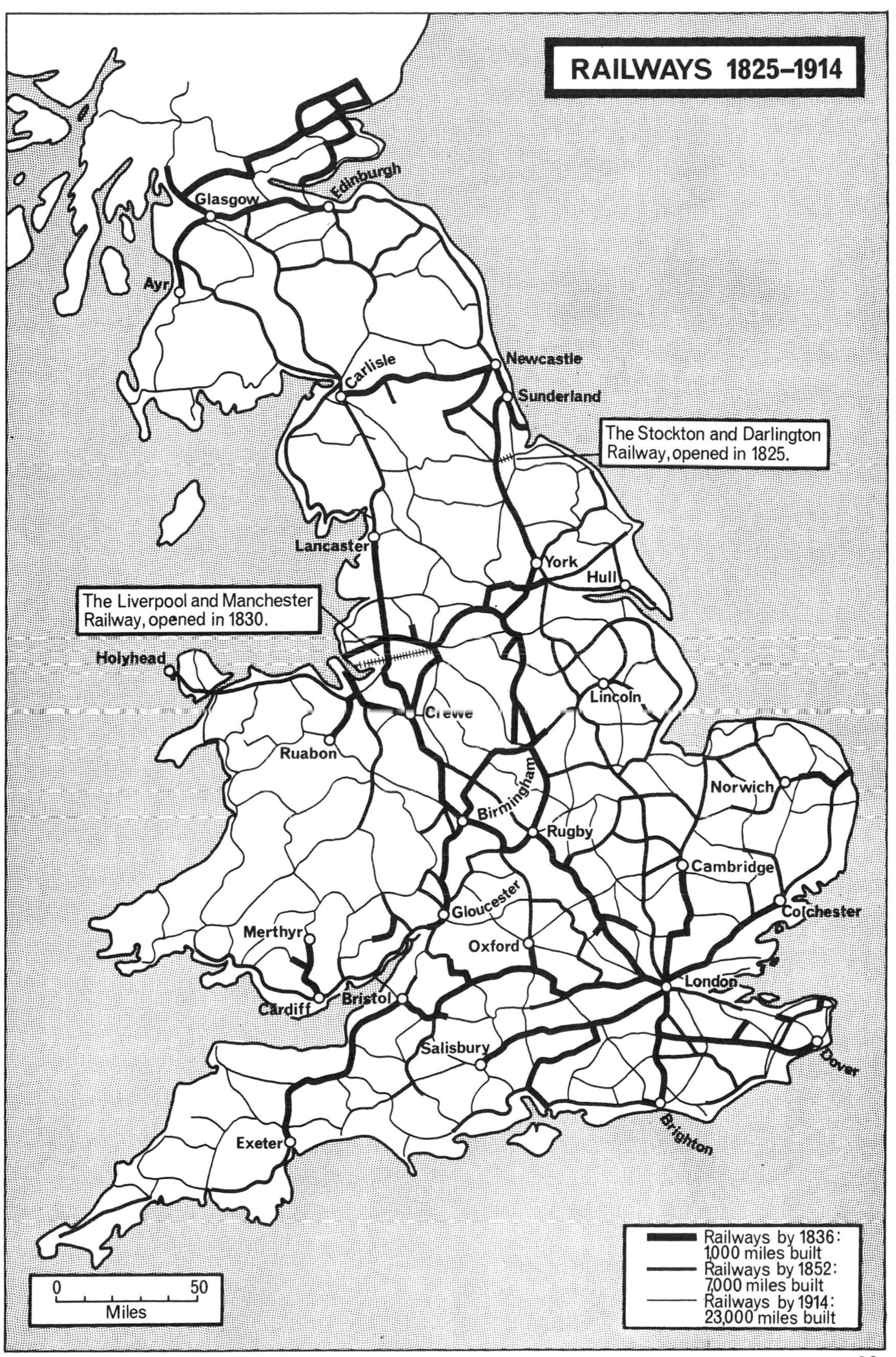
RAILWAYS 1825–1914
Glasgow
Edinburgh
Ayr
Carlisle
Newcastle
Sunderland
The Stockton and Darlington Railway, opened in 1825.
Lancaster
York
Hull
The Liverpool and Manchester Railway, opened in 1830.
Holyhead
Lincoln
Crewe
Ruabon
Birmingham
Norwich
Rugby
Cambridge
Gloucester
Colchester
Merthyr
Oxford
London
Cardiff
Bristol
Salisbury
Dover
Brighton
Exeter
Railways by 1836: 1,000 miles built
Railways by 1852: 7,000 miles built
Railways by 1914: 23,000 miles built
0
50
Miles

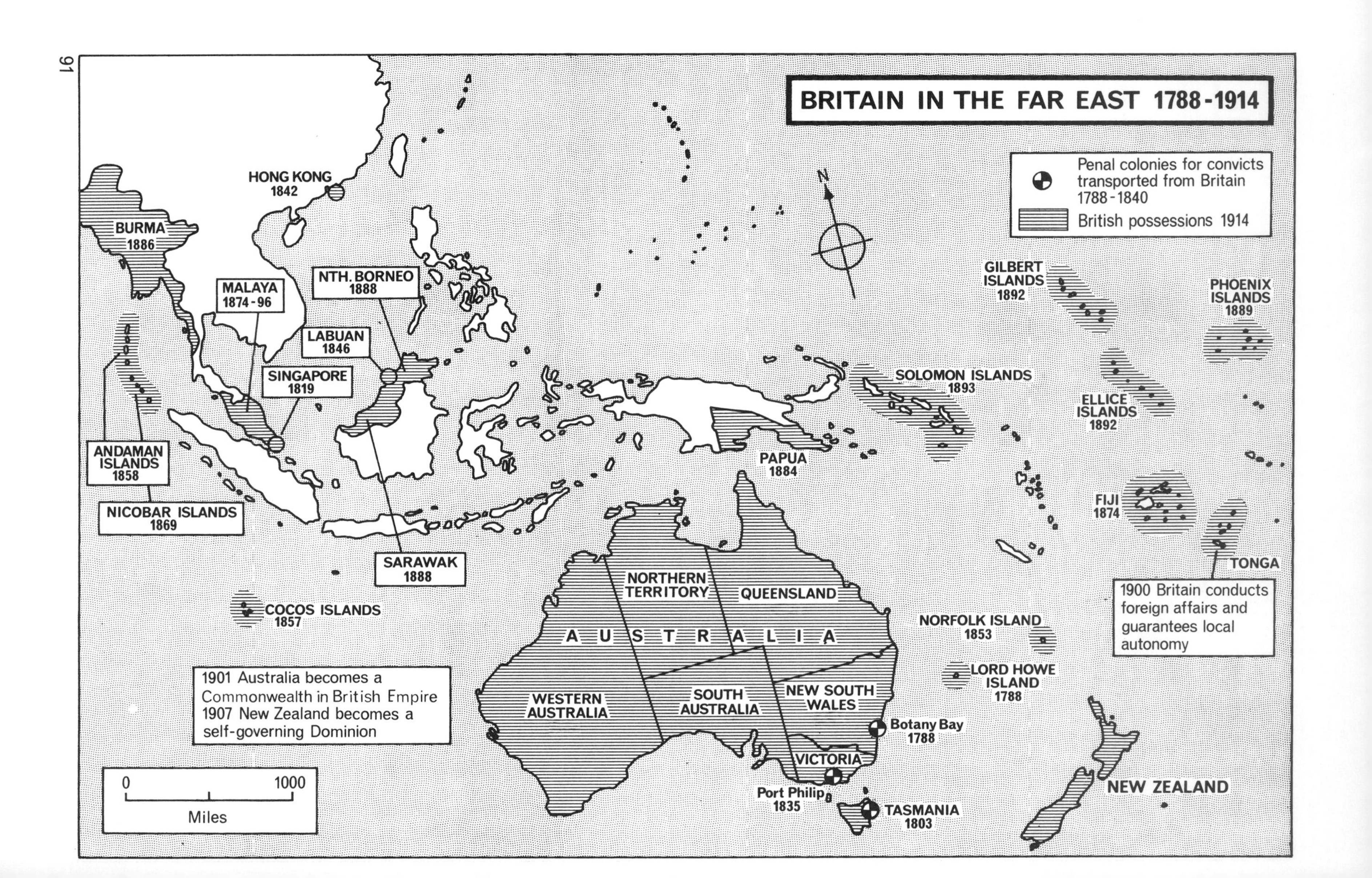
BRITAIN IN THE FAR EAST 1788-1914
Penal colonies for convicts transported from Britain 1788-1840
British possessions 1914
N
HONG KONG 1842
BURMA 1886
MALAYA 1874-96
NTH. BORNEO 1888
LABUAN 1846
SINGAPORE 1819
ANDAMAN ISLANDS 1858
NICOBAR ISLANDS 1869
SARAWAK 1888
COCOS ISLANDS 1857
GILBERT ISLANDS 1892
PHOENIX ISLANDS 1889
SOLOMON ISLANDS 1893
ELLICE ISLANDS 1892
PAPUA 1884
FIJI 1874
TONGA
1900 Britain conducts foreign affairs and guarantees local autonomy
NORTHERN TERRITORY
QUEENSLAND
A U S T R A L I A
NORFOLK ISLAND 1853
LORD HOWE ISLAND 1788
WESTERN AUSTRALIA
SOUTH AUSTRALIA
NEW SOUTH WALES
Botany Bay 1788
VICTORIA
Port Philip 1835
TASMANIA 1803
NEW ZEALAND
1901 Australia becomes a Commonwealth in British Empire
1907 New Zealand becomes a self-governing Dominion
0
1000
Miles

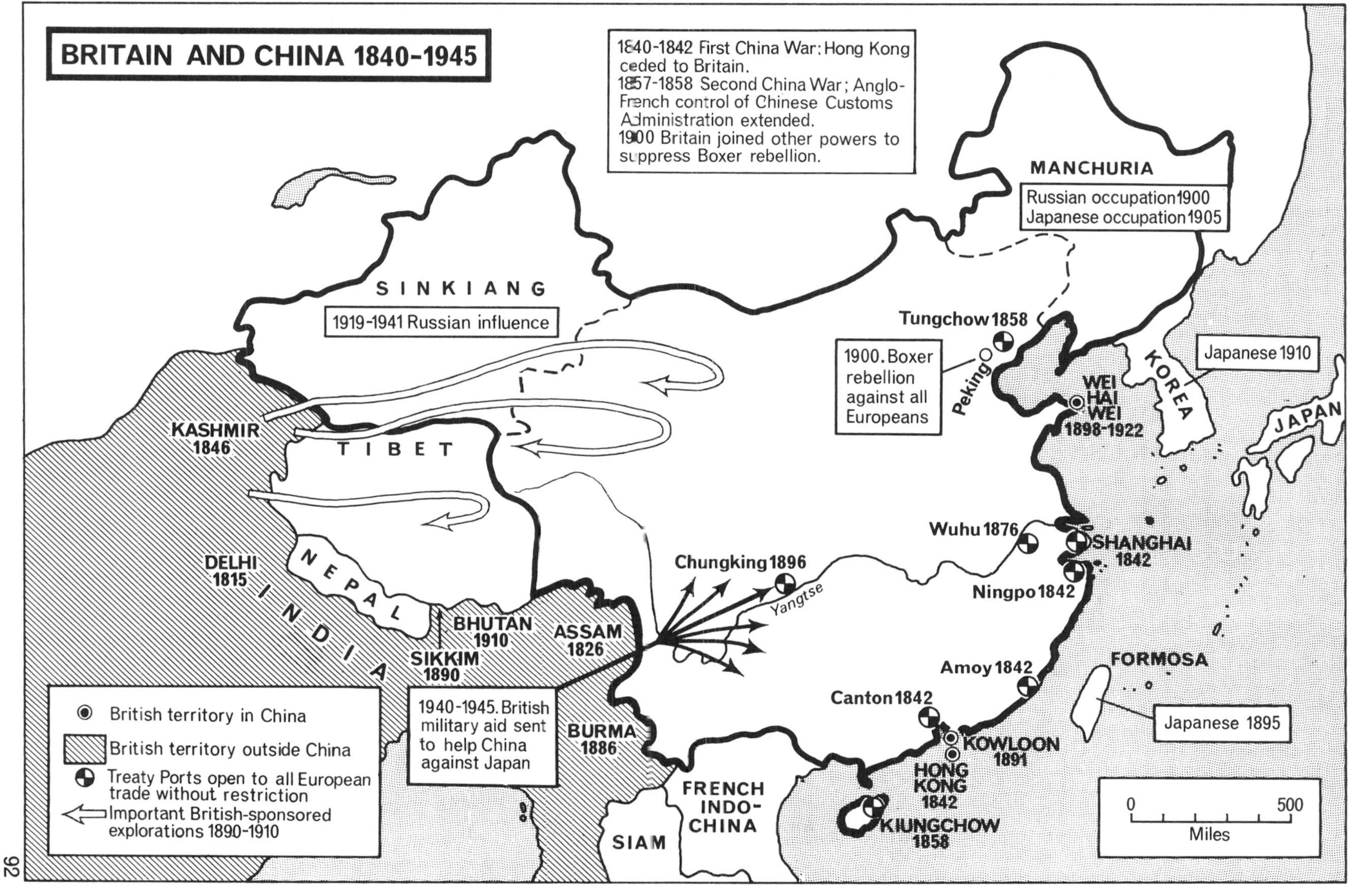
BRITAIN AND CHINA 1840-1945
1840-1842 First China War: Hong Kong ceded to Britain.
1857-1858 Second China War; Anglo-French control of Chinese Customs Administration extended.
1900 Britain joined other powers to suppress Boxer rebellion.
MANCHURIA
Russian occupation 1900
Japanese occupation 1905
SINKIANG
1919-1941 Russian influence
Tungchow 1858
Peking
1900. Boxer rebellion against all Europeans
KOREA
Japanese 1910
JAPAN
WEI HAI WEI 1898-1922
KASHMIR 1846
TIBET
Wuhu 1876
SHANGHAI 1842
Chungking 1896
Yangtse
Ningpo 1842
DELHI 1815
NEPAL
INDIA
BHUTAN 1910
SIKKIM 1890
ASSAM 1826
Amoy 1842
FORMOSA
Canton 1842
Japanese 1895
KOWLOON 1891
HONG KONG 1842
KIUNGCHOW 1858
1940-1945. British military aid sent to help China against Japan
BURMA 1886
FRENCH INDO-CHINA
SIAM
0
500
Miles
British territory in China
British territory outside China
Treaty Ports open to all European trade without restriction
Important British-sponsored explorations 1890-1910

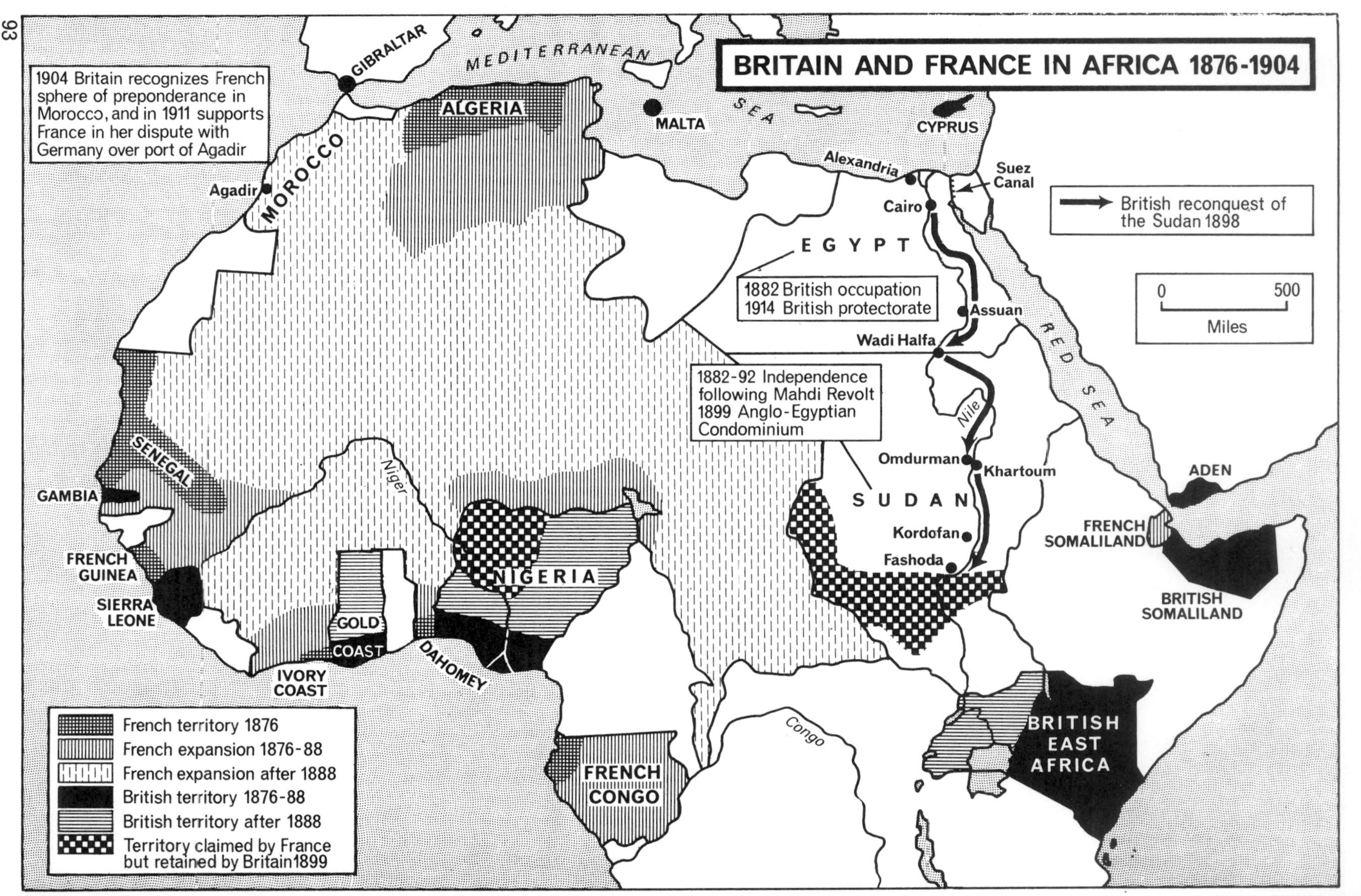
BRITAIN AND FRANCE IN AFRICA 1876-1904
1904 Britain recognizes French sphere of preponderance in Morocco, and in 1911 supports France in her dispute with Germany over port of Agadir
GIBRALTAR
MEDITERRANEAN SEA
MALTA
CYPRUS
ALGERIA
MOROCCO
Agadir
Alexandria
Suez Canal
Cairo
British reconquest of the Sudan 1898
EGYPT
1882 British occupation
1914 British protectorate
Assuan
Wadi Halfa
0 500 Miles
RED SEA
1882-92 Independence following Mahdi Revolt 1899 Anglo-Egyptian Condominium
Nile
Omdurman
Khartoum
SUDAN
Kordofan
Fashoda
ADEN
FRENCH SOMALILAND
BRITISH SOMALILAND
SENEGAL
GAMBIA
FRENCH GUINEA
SIERRA LEONE
Niger
NIGERIA
GOLD COAST
IVORY COAST
DAHOMEY
FRENCH CONGO
Congo
BRITISH EAST AFRICA
French territory 1876
French expansion 1876-88
French expansion after 1888
British territory 1876-88
British territory after 1888
Territory claimed by France but retained by Britain1899

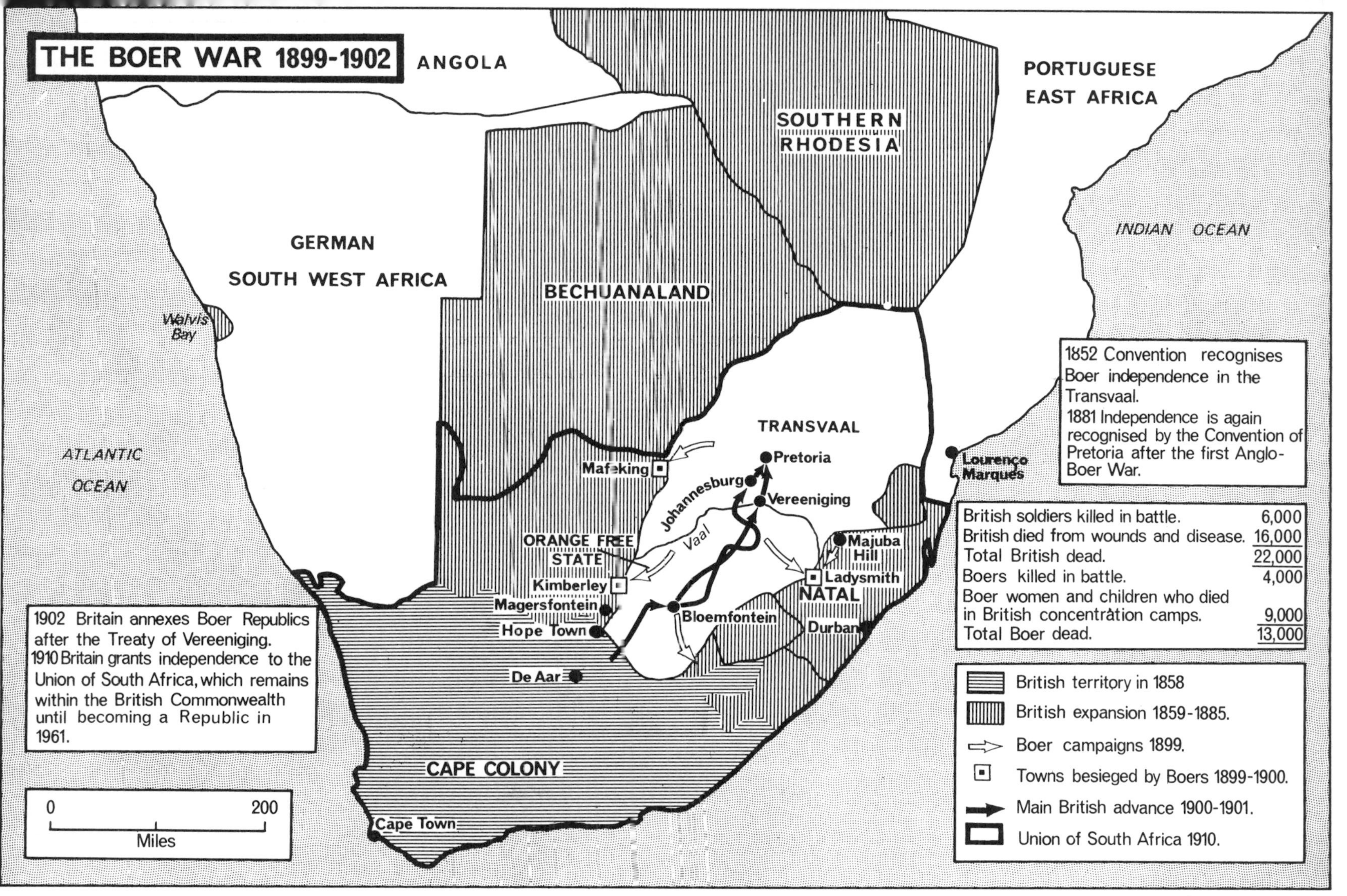
THE BOER WAR 1899-1902
ANGOLA
PORTUGUESE
EAST AFRICA
SOUTHERN
RHODESIA
GERMAN
SOUTH WEST AFRICA
BECHUANALAND
INDIAN OCEAN
ATLANTIC
OCEAN
Walvis Bay
TRANSVAAL
Pretoria
Mafeking
Johannesburg
Vereeniging
Vaal
ORANGE FREE
STATE
Kimberley
Magersfontein
Hope Town
Bloemfontein
Majuba
Hill
Ladysmith
NATAL
Durban
De Aar
CAPE COLONY
Cape Town
Lourenço
Marques
1852 Convention recognises Boer independence in the Transvaal.
1881 Independence is again recognised by the Convention of Pretoria after the first Anglo-Boer War.
British soldiers killed in battle. 6,000
British died from wounds and disease. 16,000
Total British dead. 22,000
Boers killed in battle. 4,000
Boer women and children who died in British concentration camps. 9,000
Total Boer dead. 13,000
British territory in 1858
British expansion 1859-1885.
Boer campaigns 1899.
Towns besieged by Boers 1899-1900.
Main British advance 1900-1901.
Union of South Africa 1910.
1902 Britain annexes Boer Republics after the Treaty of Vereeniging.
1910 Britain grants independence to the Union of South Africa, which remains within the British Commonwealth until becoming a Republic in 1961.
0
200
Miles

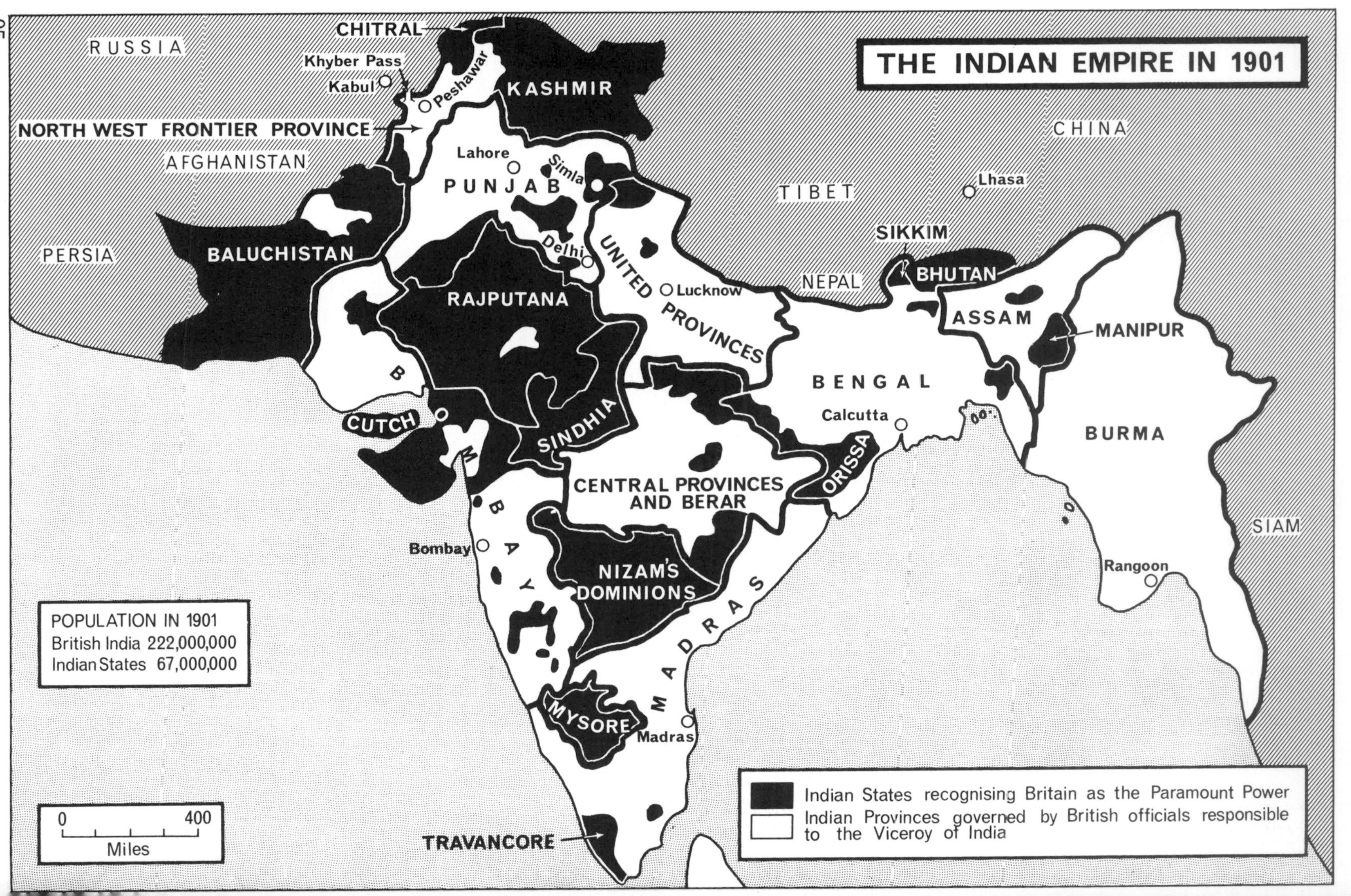
THE INDIAN EMPIRE IN 1901
RUSSIA
CHITRAL
Khyber Pass
Kabul
Peshawar
KASHMIR
NORTH WEST FRONTIER PROVINCE
AFGHANISTAN
Lahore
Simla
PUNJAB
PERSIA
BALUCHISTAN
Delhi
UNITED PROVINCES
Lucknow
RAJPUTANA
CUTCH
BOMBAY
SINDHIA
Bombay
CENTRAL PROVINCES AND BERAR
NIZAM'S DOMINIONS
MYSORE
Madras
MADRAS
TRAVANCORE
ORISSA
Calcutta
BENGAL
TIBET
NEPAL
SIKKIM
BHUTAN
Lhasa
CHINA
ASSAM
MANIPUR
BURMA
Rangoon
SIAM
POPULATION IN 1901
British India 222,000,000
Indian States 67,000,000
0
400
Miles
Indian States recognising Britain as the Paramount Power
Indian Provinces governed by British officials responsible to the Viceroy of India

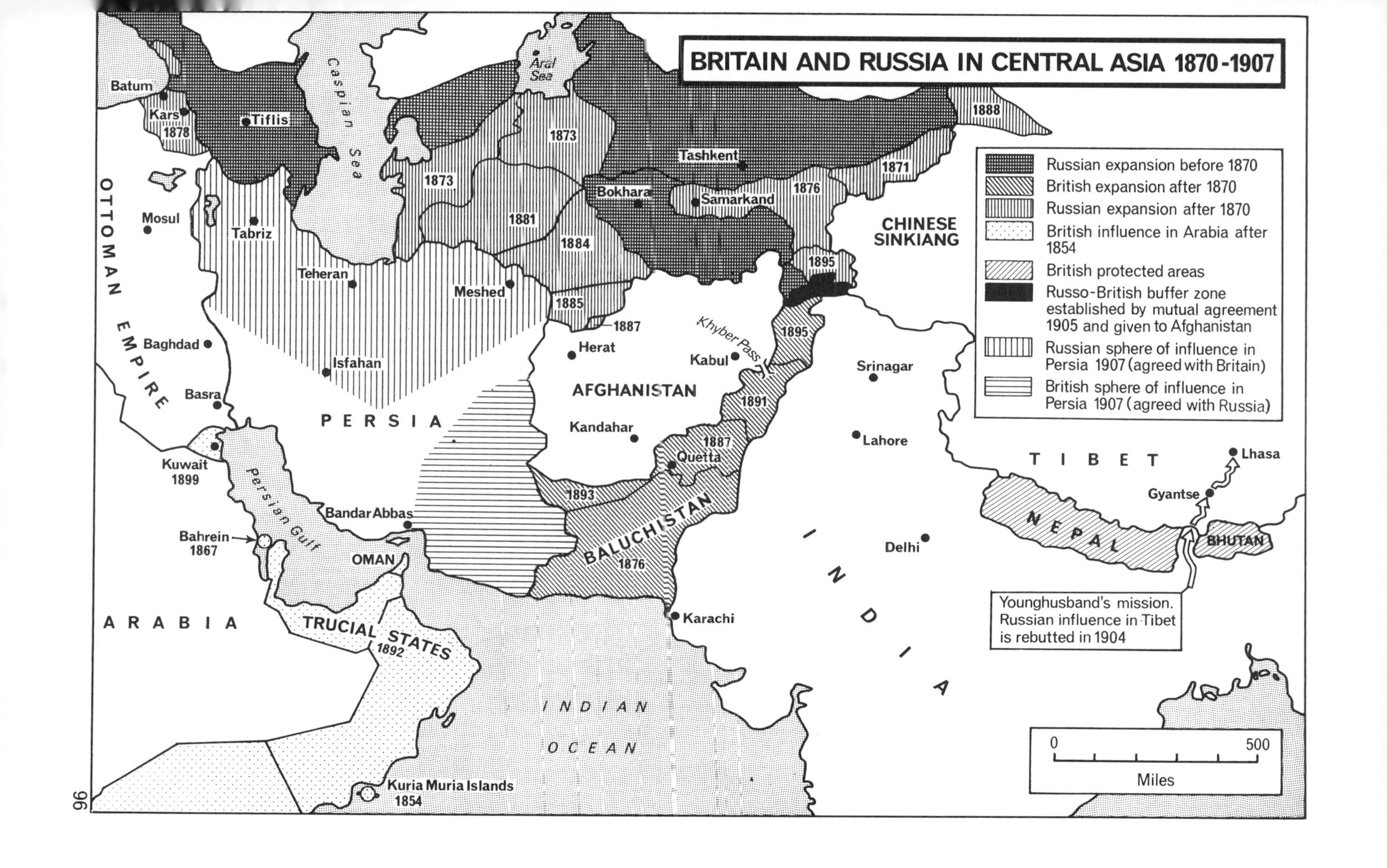
BRITAIN AND RUSSIA IN CENTRAL ASIA 1870-1907
Russian expansion before 1870
British expansion after 1870
Russian expansion after 1870
British influence in Arabia after 1854
British protected areas
Russo-British buffer zone established by mutual agreement 1905 and given to Afghanistan
Russian sphere of influence in Persia 1907 (agreed with Britain)
British sphere of influence in Persia 1907 (agreed with Russia)
Batum
Kars
1878
Tiflis
Caspian Sea
Aral Sea
1873
1873
1881
1884
1885
1887
1888
1871
1876
1895
Tashkent
Bokhara
Samarkand
CHINESE SINKIANG
OTTOMAN EMPIRE
Mosul
Tabriz
Teheran
Meshed
Baghdad
Isfahan
Basra
PERSIA
Herat
Kabul
Khyber Pass
AFGHANISTAN
Kandahar
1895
1891
1887
Quetta
1893
BALUCHISTAN
1876
Kuwait
1899
Persian Gulf
Bandar Abbas
Bahrein
1867
OMAN
ARABIA
TRUCIAL STATES
1892
INDIAN OCEAN
Kuria Muria Islands
1854
Karachi
Srinagar
Lahore
Delhi
INDIA
TIBET
Lhasa
Gyantse
NEPAL
BHUTAN
Younghusband's mission. Russian influence in Tibet is rebutted in 1904
0
500
Miles

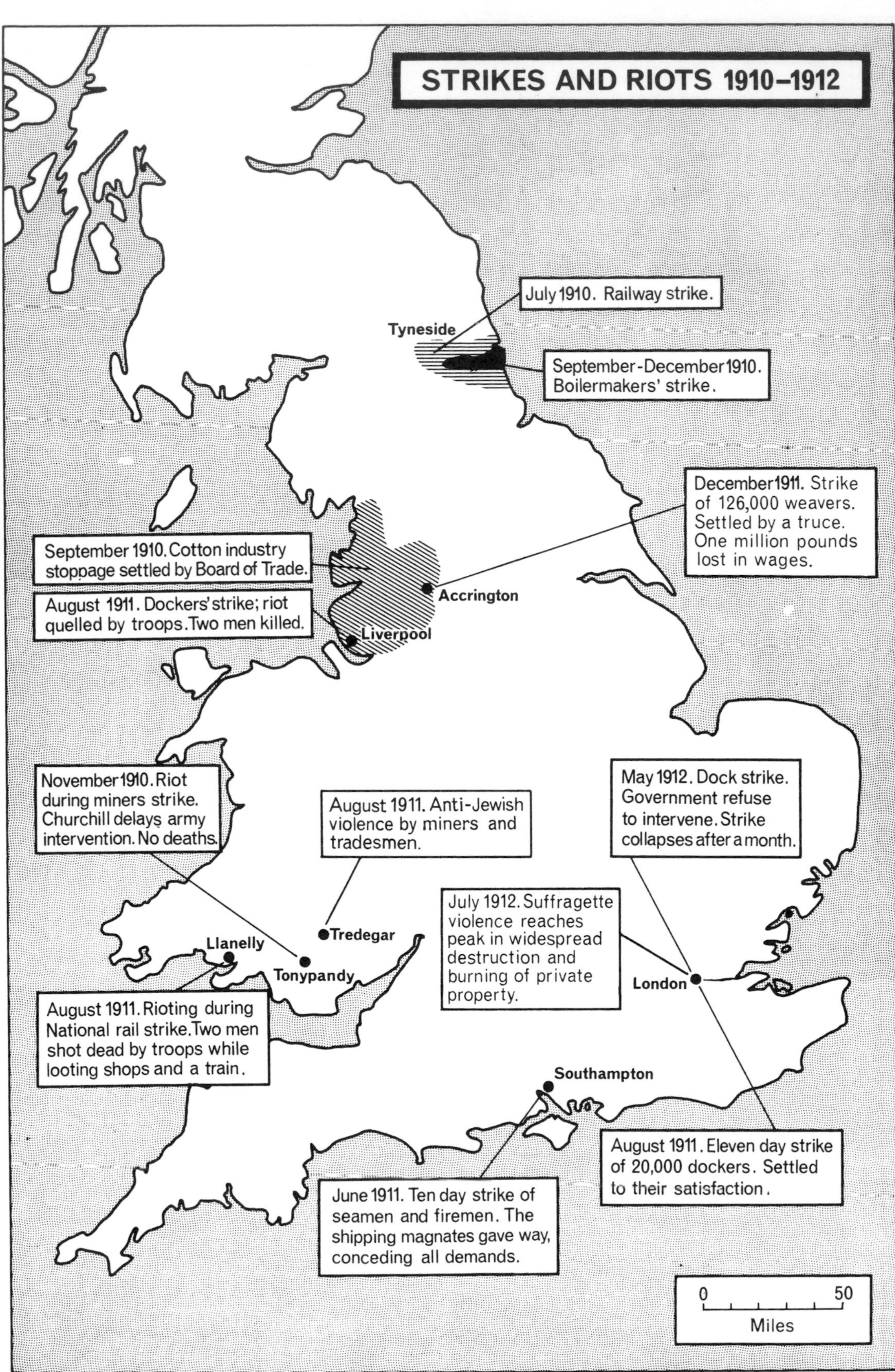

STRIKES AND RIOTS 1910–1912
July 1910. Railway strike.
Tyneside
September-December 1910. Boilermakers' strike.
December 1911. Strike of 126,000 weavers. Settled by a truce. One million pounds lost in wages.
September 1910. Cotton industry stoppage settled by Board of Trade.
August 1911. Dockers' strike; riot quelled by troops. Two men killed.
Accrington
Liverpool
November 1910. Riot during miners strike. Churchill delays army intervention. No deaths.
August 1911. Anti-Jewish violence by miners and tradesmen.
May 1912. Dock strike. Government refuse to intervene. Strike collapses after a month.
July 1912. Suffragette violence reaches peak in widespread destruction and burning of private property.
Tredegar
Llanelly
Tonypandy
London
August 1911. Rioting during National rail strike. Two men shot dead by troops while looting shops and a train.
Southampton
August 1911. Eleven day strike of 20,000 dockers. Settled to their satisfaction.
June 1911. Ten day strike of seamen and firemen. The shipping magnates gave way, conceding all demands.
0
50
Miles

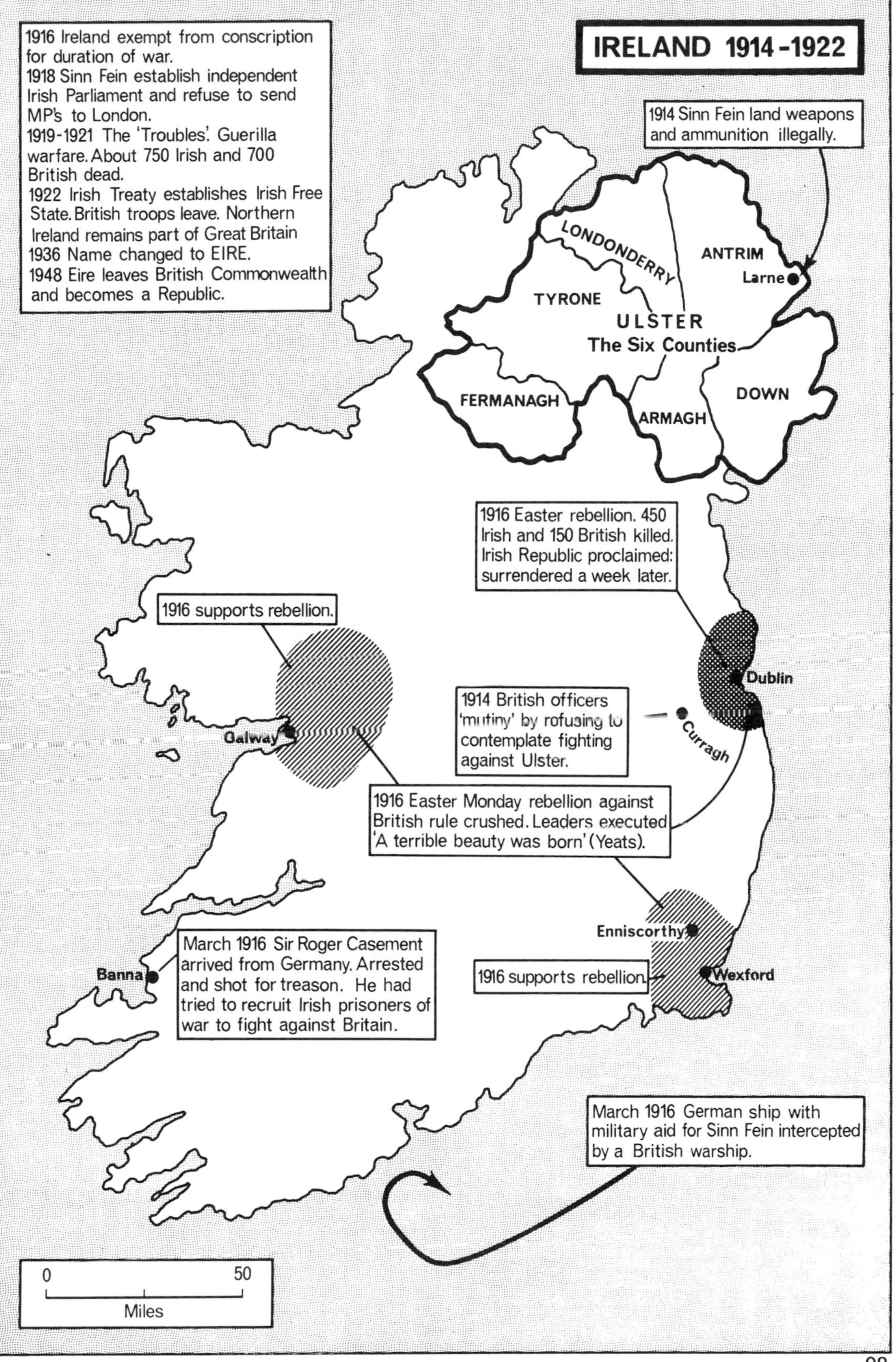
IRELAND 1914-1922
1916 Ireland exempt from conscription for duration of war.
1918 Sinn Fein establish independent Irish Parliament and refuse to send MP's to London.
1919-1921 The 'Troubles'. Guerilla warfare. About 750 Irish and 700 British dead.
1922 Irish Treaty establishes Irish Free State. British troops leave. Northern Ireland remains part of Great Britain
1936 Name changed to EIRE.
1948 Eire leaves British Commonwealth and becomes a Republic.
1914 Sinn Fein land weapons and ammunition illegally.
LONDONDERRY
ANTRIM
Larne
TYRONE
ULSTER
The Six Counties
FERMANAGH
ARMAGH
DOWN
1916 Easter rebellion. 450 Irish and 150 British killed. Irish Republic proclaimed: surrendered a week later.
1916 supports rebellion.
Dublin
1914 British officers 'mutiny' by refusing to contemplate fighting against Ulster.
Curragh
Galway
1916 Easter Monday rebellion against British rule crushed. Leaders executed 'A terrible beauty was born' (Yeats).
Enniscorthy
Banna
March 1916 Sir Roger Casement arrived from Germany. Arrested and shot for treason. He had tried to recruit Irish prisoners of war to fight against Britain.
1916 supports rebellion.
Wexford
March 1916 German ship with military aid for Sinn Fein intercepted by a British warship.
0
50
Miles

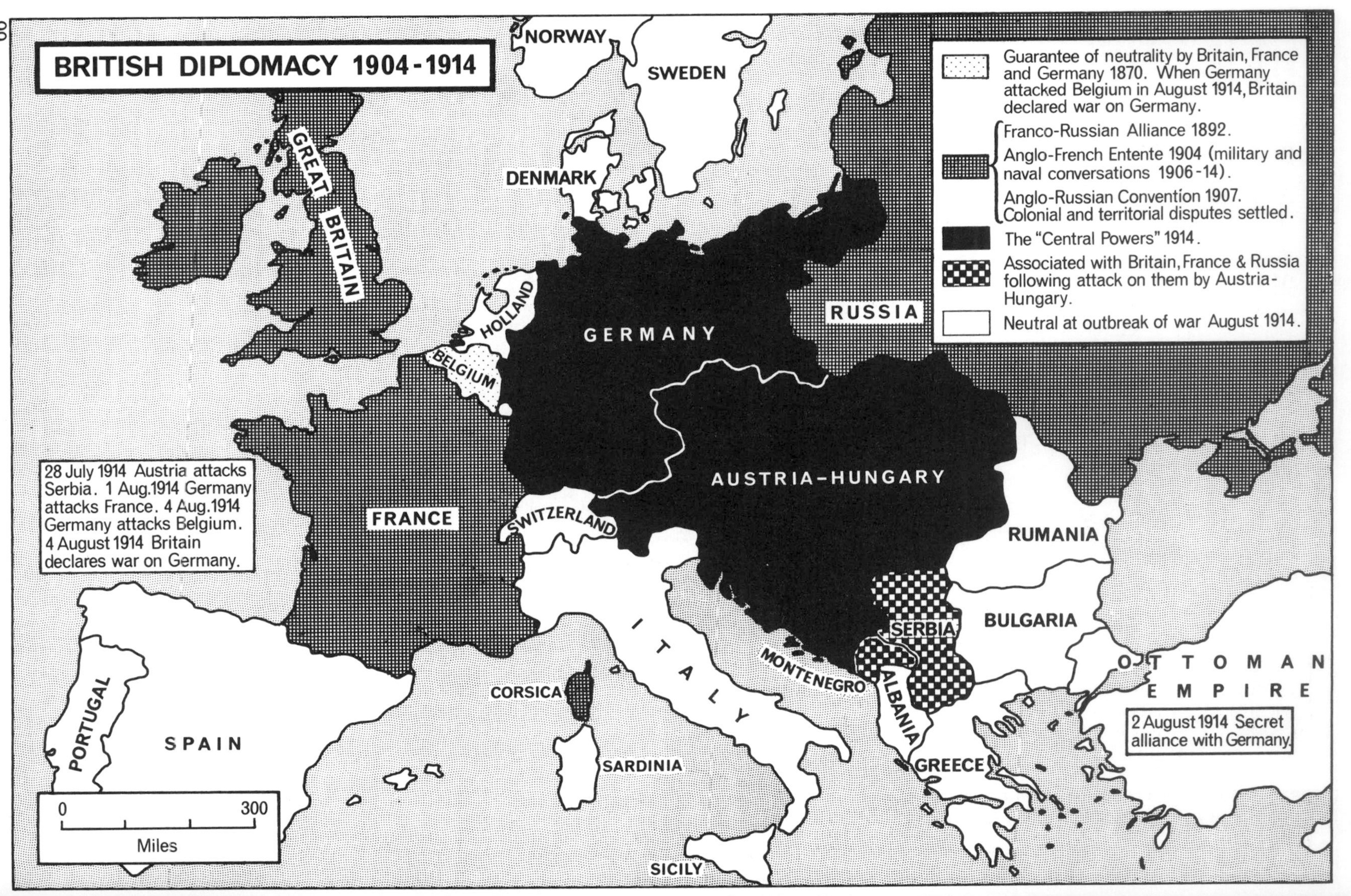
BRITISH DIPLOMACY 1904-1914
Guarantee of neutrality by Britain, France and Germany 1870. When Germany attacked Belgium in August 1914, Britain declared war on Germany.
Franco-Russian Alliance 1892.
Anglo-French Entente 1904 (military and naval conversations 1906-14).
Anglo-Russian Convention 1907. Colonial and territorial disputes settled.
The "Central Powers" 1914.
Associated with Britain, France & Russia following attack on them by Austria-Hungary.
Neutral at outbreak of war August 1914.
28 July 1914 Austria attacks Serbia. 1 Aug.1914 Germany attacks France. 4 Aug.1914 Germany attacks Belgium. 4 August 1914 Britain declares war on Germany.
2 August 1914 Secret alliance with Germany.
NORWAY
SWEDEN
DENMARK
GREAT BRITAIN
HOLLAND
BELGIUM
GERMANY
RUSSIA
FRANCE
SWITZERLAND
AUSTRIA-HUNGARY
RUMANIA
BULGARIA
SERBIA
MONTENEGRO
ALBANIA
GREECE
OTTOMAN EMPIRE
ITALY
CORSICA
SARDINIA
SICILY
PORTUGAL
SPAIN
0
300
Miles

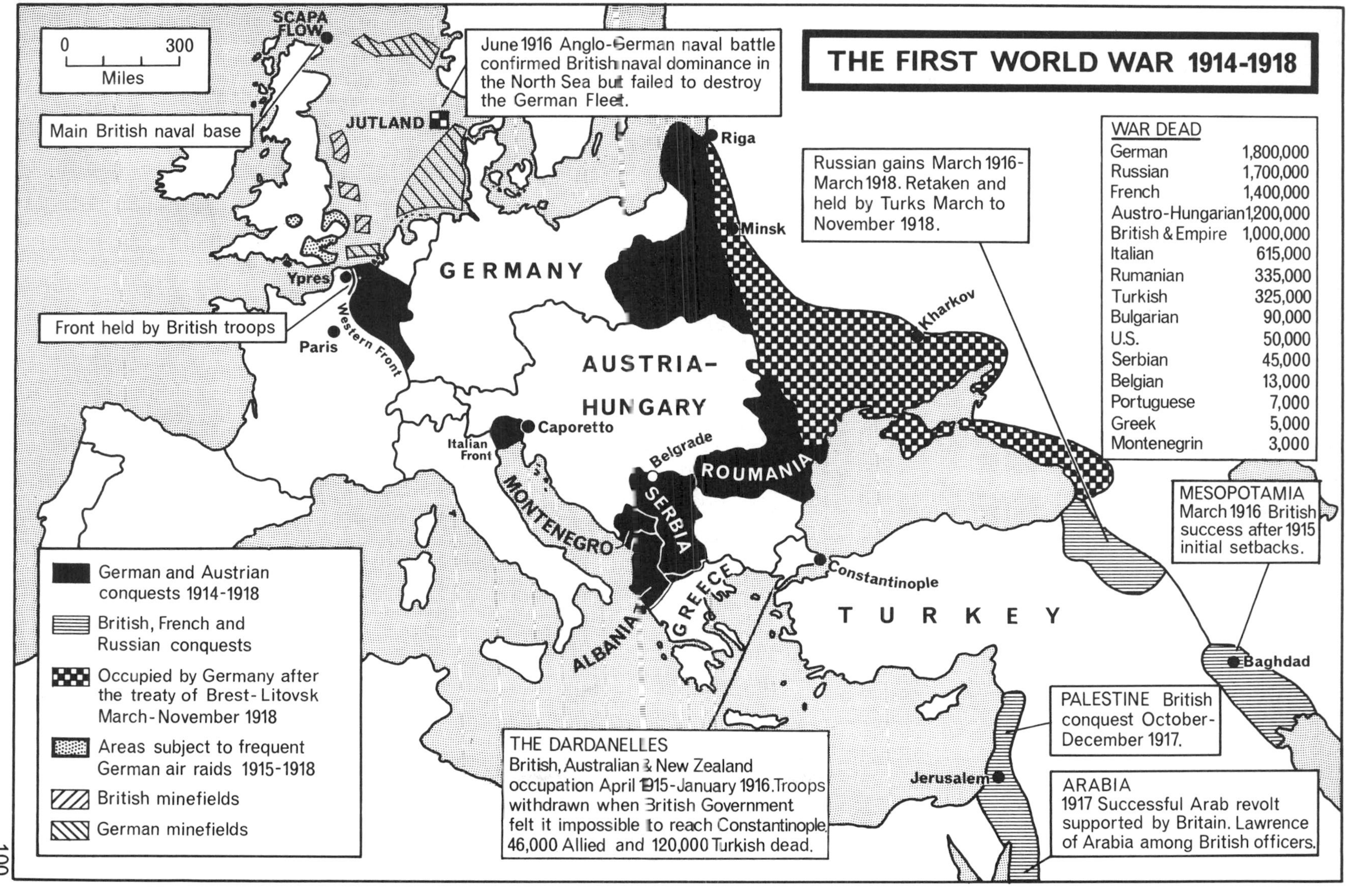

WAR DEAD	
German	1,800,000
Russian	1,700,000
French	1,400,000
Austro-Hungarian	1,200,000
British & Empire	1,000,000
Italian	615,000
Rumanian	335,000
Turkish	325,000
Bulgarian	90,000
U.S.	50,000
Serbian	45,000
Belgian	13,000
Portuguese	7,000
Greek	5,000
Montenegrin	3,000

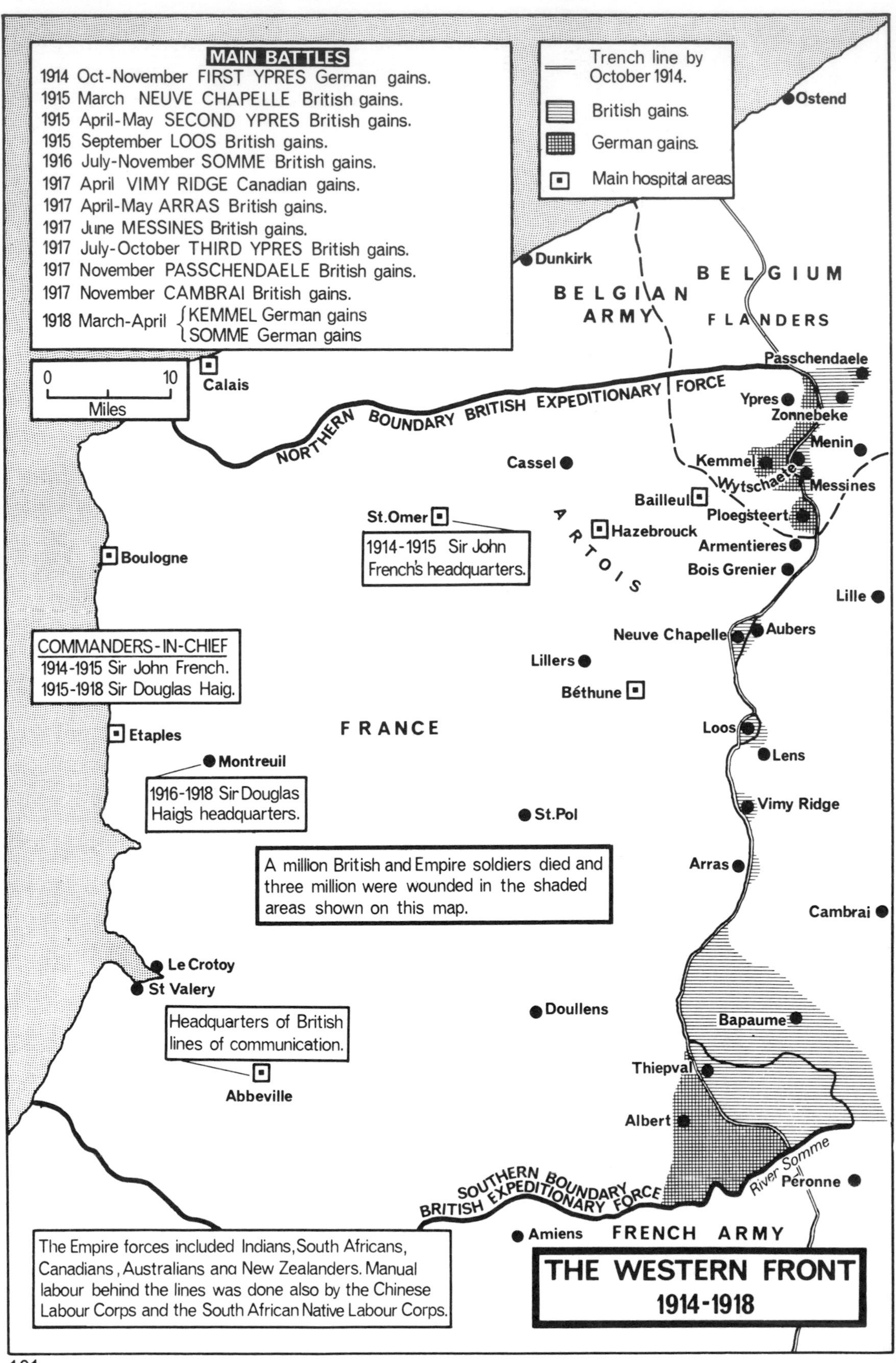
MAIN BATTLES
1914 Oct-November FIRST YPRES German gains.
1915 March NEUVE CHAPELLE British gains.
1915 April-May SECOND YPRES British gains.
1915 September LOOS British gains.
1916 July-November SOMME British gains.
1917 April VIMY RIDGE Canadian gains.
1917 April-May ARRAS British gains.
1917 June MESSINES British gains.
1917 July-October THIRD YPRES British gains.
1917 November PASSCHENDAELE British gains.
1917 November CAMBRAI British gains.
1918 March-April KEMMEL German gains
SOMME German gains
Trench line by October 1914.
British gains.
German gains.
Main hospital areas.
0 10 Miles
Ostend
Dunkirk
Calais
BELGIAN ARMY
BELGIUM
FLANDERS
Passchendaele
Ypres
Zonnebeke
Menin
Kemmel
Wytschaete
Messines
NORTHERN BOUNDARY BRITISH EXPEDITIONARY FORCE
Cassel
Bailleul
Ploegsteert
St.Omer
1914-1915 Sir John French's headquarters.
Hazebrouck
ARTOIS
Armentieres
Bois Grenier
Lille
Boulogne
Neuve Chapelle
Aubers
Lillers
Béthune
COMMANDERS-IN-CHIEF
1914-1915 Sir John French.
1915-1918 Sir Douglas Haig.
Etaples
FRANCE
Loos
Lens
Montreuil
1916-1918 Sir Douglas Haig's headquarters.
St.Pol
Vimy Ridge
Arras
A million British and Empire soldiers died and three million were wounded in the shaded areas shown on this map.
Cambrai
Le Crotoy
St Valery
Doullens
Headquarters of British lines of communication.
Abbeville
Bapaume
Thiepval
Albert
River Somme
Peronne
SOUTHERN BOUNDARY BRITISH EXPEDITIONARY FORCE
Amiens
FRENCH ARMY
The Empire forces included Indians, South Africans, Canadians, Australians and New Zealanders. Manual labour behind the lines was done also by the Chinese Labour Corps and the South African Native Labour Corps.
THE WESTERN FRONT 1914-1918

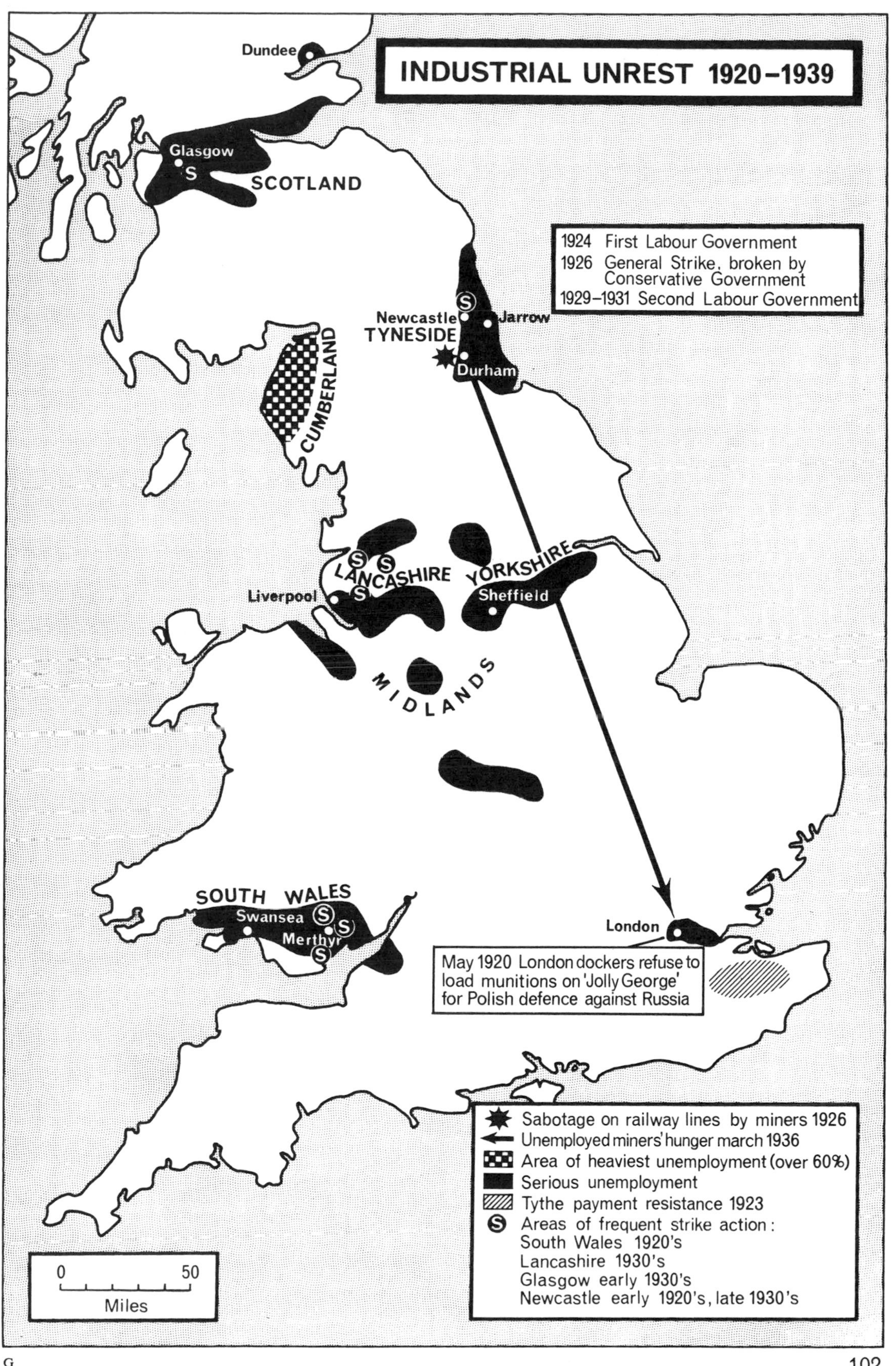
INDUSTRIAL UNREST 1920–1939
Dundee
Glasgow
S
SCOTLAND
1924 First Labour Government
1926 General Strike, broken by Conservative Government
1929–1931 Second Labour Government
Newcastle
Jarrow
TYNESIDE
Durham
CUMBERLAND
LANCASHIRE
YORKSHIRE
Liverpool
Sheffield
MIDLANDS
SOUTH WALES
Swansea
Merthyr
London
May 1920 London dockers refuse to load munitions on 'Jolly George' for Polish defence against Russia
Sabotage on railway lines by miners 1926
Unemployed miners' hunger march 1936
Area of heaviest unemployment (over 60%)
Serious unemployment
Tythe payment resistance 1923
Areas of frequent strike action:
South Wales 1920's
Lancashire 1930's
Glasgow early 1930's
Newcastle early 1920's, late 1930's
0
50
Miles

G

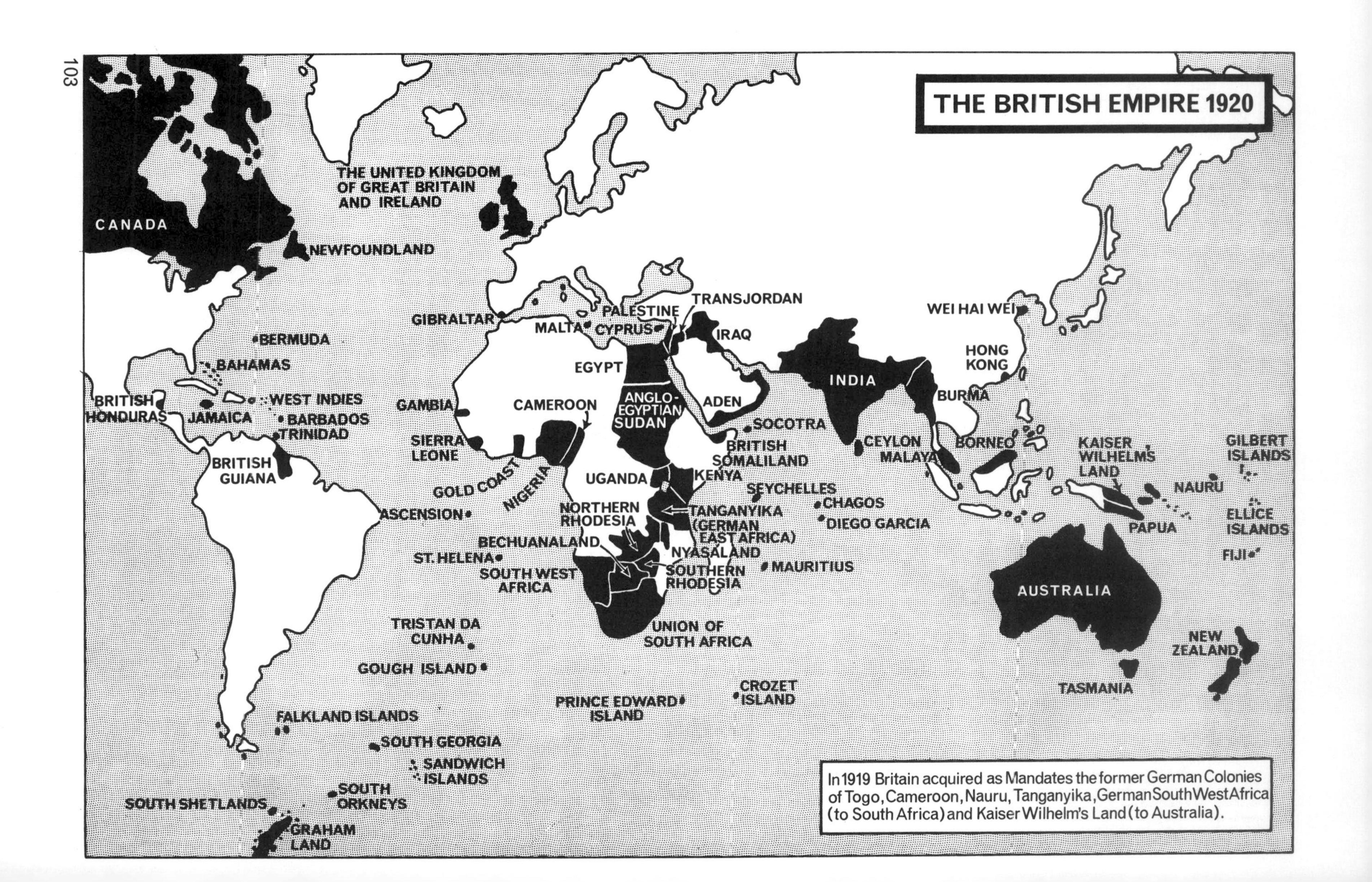
THE BRITISH EMPIRE 1920
CANADA
THE UNITED KINGDOM OF GREAT BRITAIN AND IRELAND
NEWFOUNDLAND
BERMUDA
BAHAMAS
BRITISH HONDURAS
JAMAICA
WEST INDIES
BARBADOS
TRINIDAD
BRITISH GUIANA
GIBRALTAR
MALTA
PALESTINE
CYPRUS
TRANSJORDAN
IRAQ
EGYPT
GAMBIA
SIERRA LEONE
GOLD COAST
NIGERIA
CAMEROON
ANGLO-EGYPTIAN SUDAN
ADEN
SOCOTRA
BRITISH SOMALILAND
UGANDA
KENYA
SEYCHELLES
TANGANYIKA (GERMAN EAST AFRICA)
NORTHERN RHODESIA
BECHUANALAND
NYASALAND
SOUTHERN RHODESIA
SOUTH WEST AFRICA
UNION OF SOUTH AFRICA
ASCENSION
ST. HELENA
TRISTAN DA CUNHA
GOUGH ISLAND
PRINCE EDWARD ISLAND
CROZET ISLAND
FALKLAND ISLANDS
SOUTH GEORGIA
SANDWICH ISLANDS
SOUTH ORKNEYS
SOUTH SHETLANDS
GRAHAM LAND
INDIA
CEYLON
CHAGOS
DIEGO GARCIA
MAURITIUS
WEI HAI WEI
HONG KONG
BURMA
MALAYA
BORNEO
KAISER WILHELMS LAND
PAPUA
NAURU
GILBERT ISLANDS
ELLICE ISLANDS
FIJI
AUSTRALIA
TASMANIA
NEW ZEALAND
In 1919 Britain acquired as Mandates the former German Colonies of Togo, Cameroon, Nauru, Tanganyika, German South West Africa (to South Africa) and Kaiser Wilhelm's Land (to Australia).

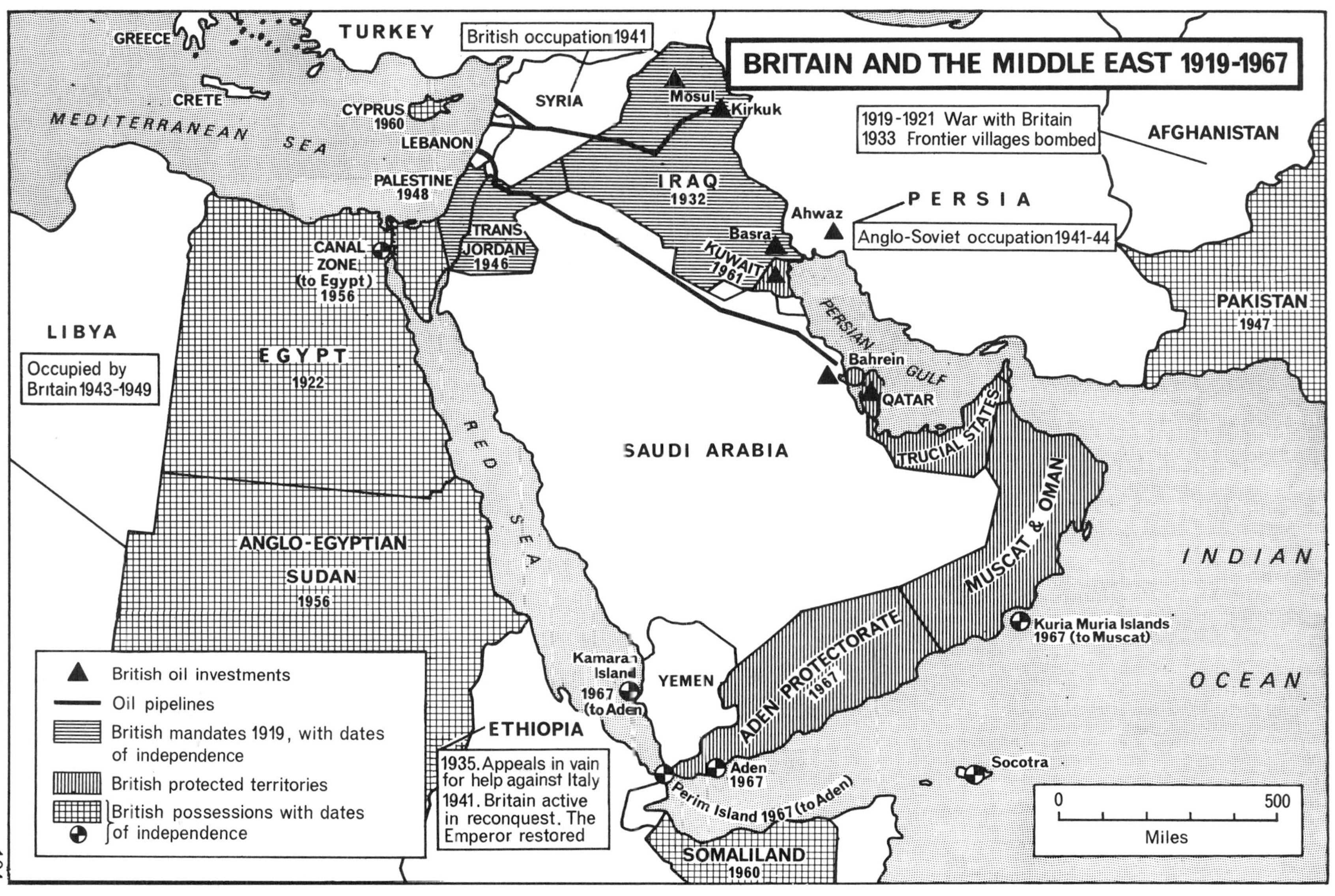
BRITAIN AND THE MIDDLE EAST 1919-1967
GREECE
TURKEY
British occupation 1941
SYRIA
Mosul
Kirkuk
1919-1921 War with Britain
1933 Frontier villages bombed
AFGHANISTAN
CRETE
MEDITERRANEAN SEA
CYPRUS 1960
LEBANON
PALESTINE 1948
IRAQ 1932
PERSIA
Ahwaz
Anglo-Soviet occupation 1941-44
Basra
KUWAIT 1961
TRANS JORDAN 1946
CANAL ZONE (to Egypt) 1956
PAKISTAN 1947
LIBYA
Occupied by Britain 1943-1949
EGYPT 1922
PERSIAN GULF
Bahrein
QATAR
TRUCIAL STATES
SAUDI ARABIA
RED SEA
MUSCAT & OMAN
ANGLO-EGYPTIAN SUDAN 1956
INDIAN OCEAN
ADEN PROTECTORATE 1967
Kuria Muria Islands 1967 (to Muscat)
Kamaran Island 1967 (to Aden)
YEMEN
ETHIOPIA
1935. Appeals in vain for help against Italy
1941. Britain active in reconquest. The Emperor restored
Aden 1967
Perim Island 1967 (to Aden)
Socotra
SOMALILAND 1960
0 500 Miles
British oil investments
Oil pipelines
British mandates 1919, with dates of independence
British protected territories
British possessions with dates of independence

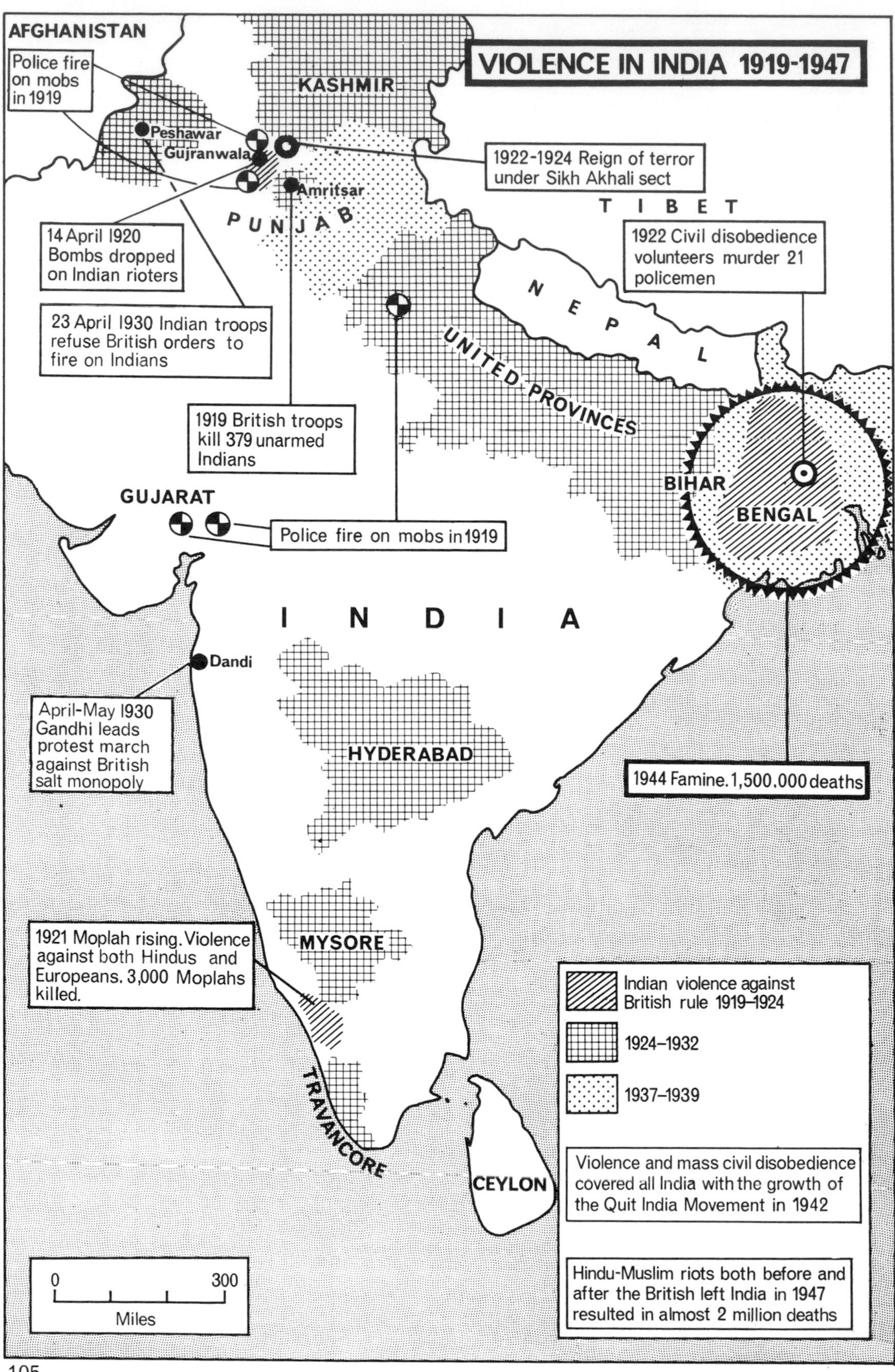

VIOLENCE IN INDIA 1919-1947
AFGHANISTAN
KASHMIR
Police fire on mobs in 1919
Peshawar
Gujranwala
Amritsar
1922-1924 Reign of terror under Sikh Akhali sect
T I B E T
PUNJAB
14 April 1920 Bombs dropped on Indian rioters
1922 Civil disobedience volunteers murder 21 policemen
NEPAL
23 April 1930 Indian troops refuse British orders to fire on Indians
UNITED PROVINCES
1919 British troops kill 379 unarmed Indians
BIHAR
BENGAL
GUJARAT
Police fire on mobs in 1919
I N D I A
Dandi
April-May 1930 Gandhi leads protest march against British salt monopoly
HYDERABAD
1944 Famine. 1,500,000 deaths
MYSORE
1921 Moplah rising. Violence against both Hindus and Europeans. 3,000 Moplahs killed.
Indian violence against British rule 1919–1924
1924–1932
1937–1939
TRAVANCORE
CEYLON
Violence and mass civil disobedience covered all India with the growth of the Quit India Movement in 1942
0
300
Miles
Hindu-Muslim riots both before and after the British left India in 1947 resulted in almost 2 million deaths

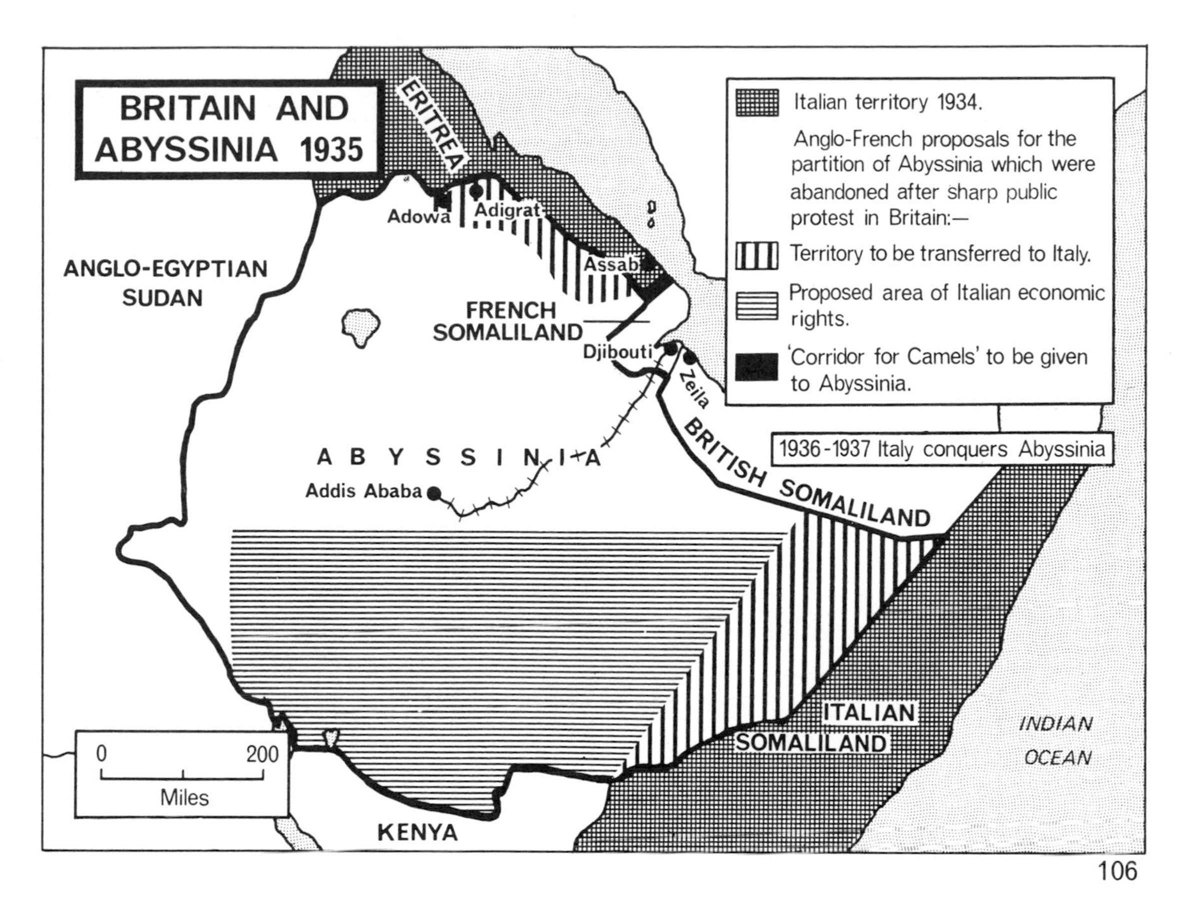
BRITAIN AND ABYSSINIA 1935
Italian territory 1934.
Anglo-French proposals for the partition of Abyssinia which were abandoned after sharp public protest in Britain:—
Territory to be transferred to Italy.
Proposed area of Italian economic rights.
'Corridor for Camels' to be given to Abyssinia.
1936-1937 Italy conquers Abyssinia
ERITREA
Adowa
Adigrat
Assab
ANGLO-EGYPTIAN SUDAN
FRENCH SOMALILAND
Djibouti
Zeila
BRITISH SOMALILAND
ABYSSINIA
Addis Ababa
ITALIAN SOMALILAND
INDIAN OCEAN
KENYA
0
200
Miles

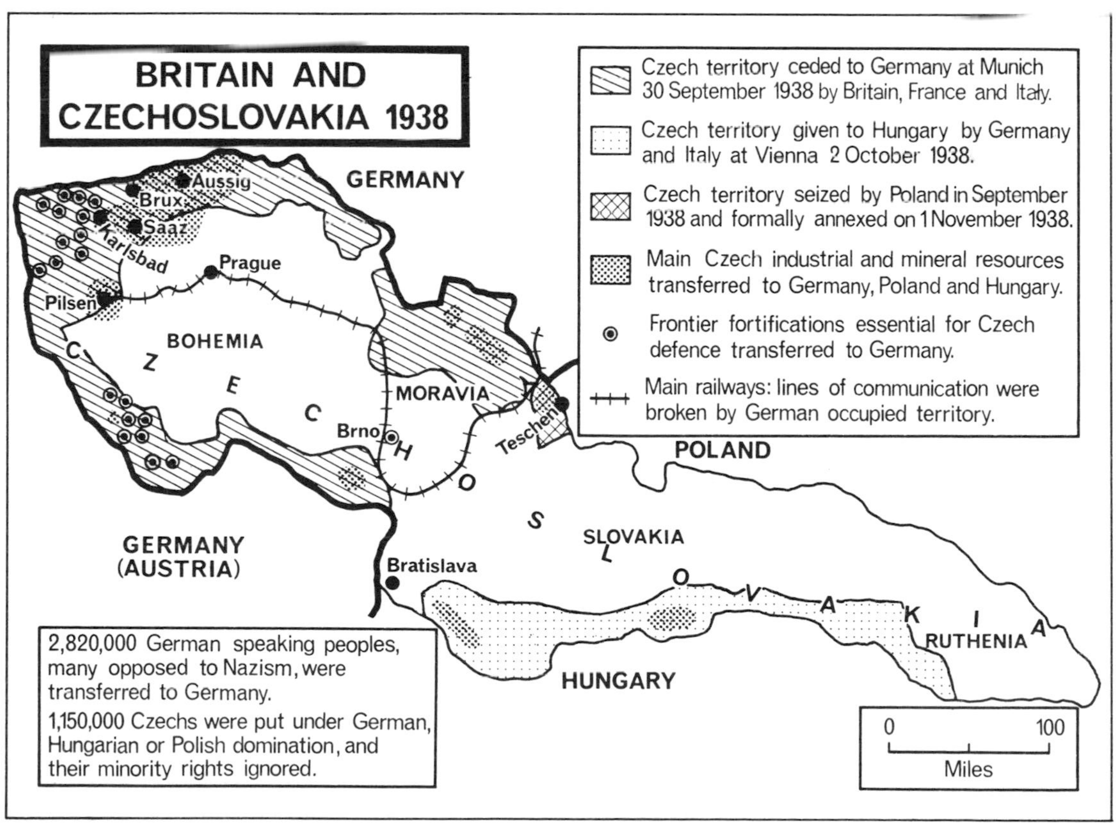
BRITAIN AND CZECHOSLOVAKIA 1938
Czech territory ceded to Germany at Munich 30 September 1938 by Britain, France and Italy.
Czech territory given to Hungary by Germany and Italy at Vienna 2 October 1938.
Czech territory seized by Poland in September 1938 and formally annexed on 1 November 1938.
Main Czech industrial and mineral resources transferred to Germany, Poland and Hungary.
Frontier fortifications essential for Czech defence transferred to Germany.
Main railways: lines of communication were broken by German occupied territory.
GERMANY
Aussig
Brux
Saaz
Karlsbad
Prague
Pilsen
BOHEMIA
MORAVIA
Brno
Teschen
POLAND
CZECHOSLOVAKIA
SLOVAKIA
RUTHENIA
GERMANY (AUSTRIA)
Bratislava
HUNGARY
2,820,000 German speaking peoples, many opposed to Nazism, were transferred to Germany.
1,150,000 Czechs were put under German, Hungarian or Polish domination, and their minority rights ignored.
0
100
Miles

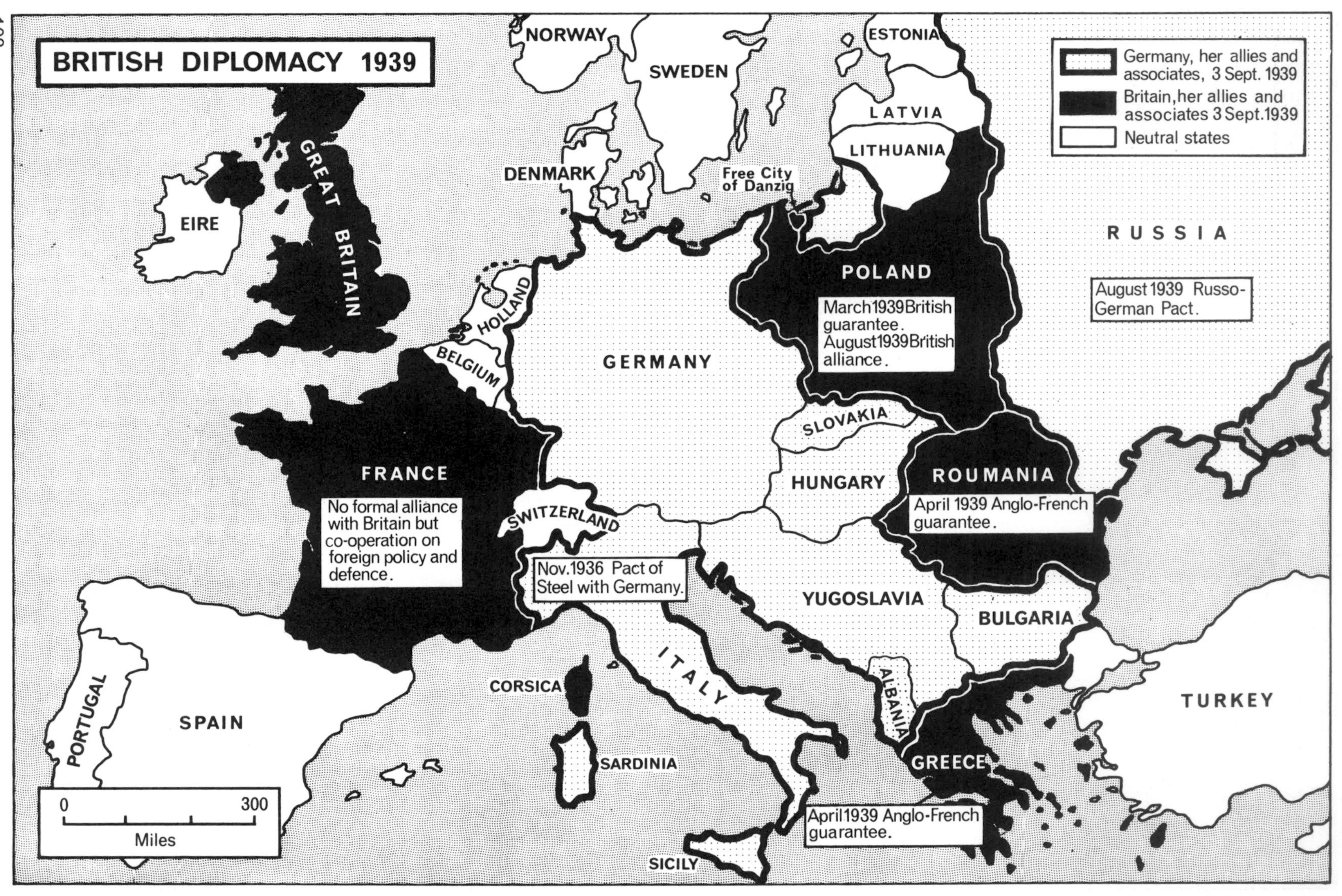

BRITISH DIPLOMACY 1939
NORWAY
SWEDEN
ESTONIA
Germany, her allies and associates, 3 Sept. 1939
Britain, her allies and associates 3 Sept. 1939
Neutral states
LATVIA
LITHUANIA
DENMARK
Free City of Danzig
GREAT BRITAIN
EIRE
RUSSIA
POLAND
March 1939 British guarantee. August 1939 British alliance.
August 1939 Russo-German Pact.
HOLLAND
BELGIUM
GERMANY
SLOVAKIA
FRANCE
HUNGARY
ROUMANIA
No formal alliance with Britain but co-operation on foreign policy and defence.
SWITZERLAND
April 1939 Anglo-French guarantee.
Nov. 1936 Pact of Steel with Germany.
YUGOSLAVIA
BULGARIA
ITALY
CORSICA
ALBANIA
TURKEY
PORTUGAL
SPAIN
SARDINIA
GREECE
0
300
Miles
April 1939 Anglo-French guarantee.
SICILY

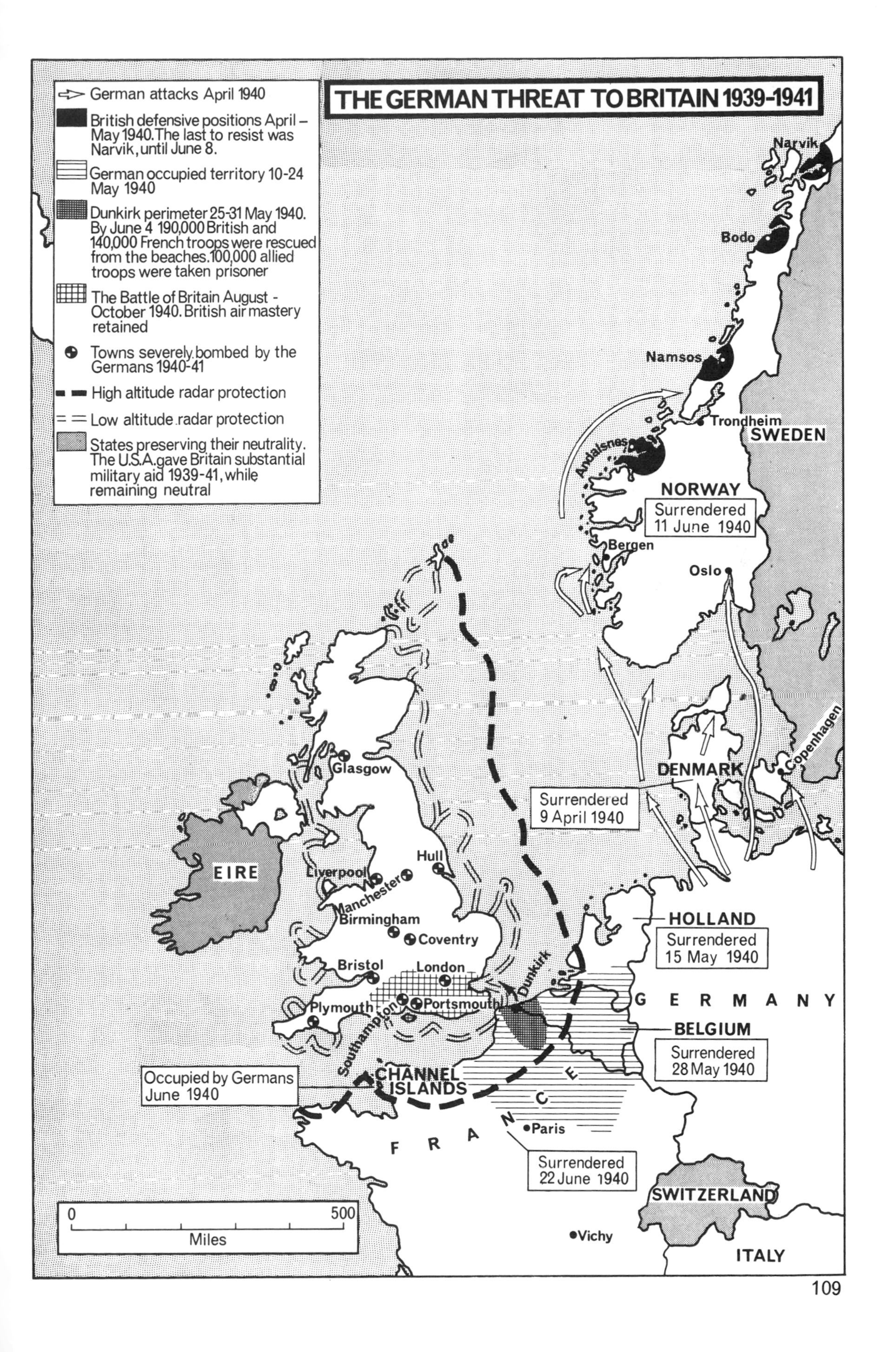
THE GERMAN THREAT TO BRITAIN 1939-1941
German attacks April 1940
British defensive positions April – May 1940. The last to resist was Narvik, until June 8.
German occupied territory 10-24 May 1940
Dunkirk perimeter 25-31 May 1940. By June 4 190,000 British and 140,000 French troops were rescued from the beaches. 100,000 allied troops were taken prisoner
The Battle of Britain August - October 1940. British air mastery retained
Towns severely bombed by the Germans 1940-41
High altitude radar protection
Low altitude radar protection
States preserving their neutrality. The U.S.A. gave Britain substantial military aid 1939-41, while remaining neutral
Narvik
Bodo
Namsos
Trondheim
SWEDEN
Andalsnes
NORWAY
Surrendered 11 June 1940
Bergen
Oslo
Copenhagen
DENMARK
Surrendered 9 April 1940
Glasgow
EIRE
Hull
Liverpool
Manchester
Birmingham
Coventry
HOLLAND
Surrendered 15 May 1940
Bristol
London
Dunkirk
GERMANY
Plymouth
Portsmouth
Southampton
BELGIUM
Surrendered 28 May 1940
Occupied by Germans June 1940
CHANNEL ISLANDS
FRANCE
Paris
Surrendered 22 June 1940
SWITZERLAND
0
500
Miles
Vichy
ITALY

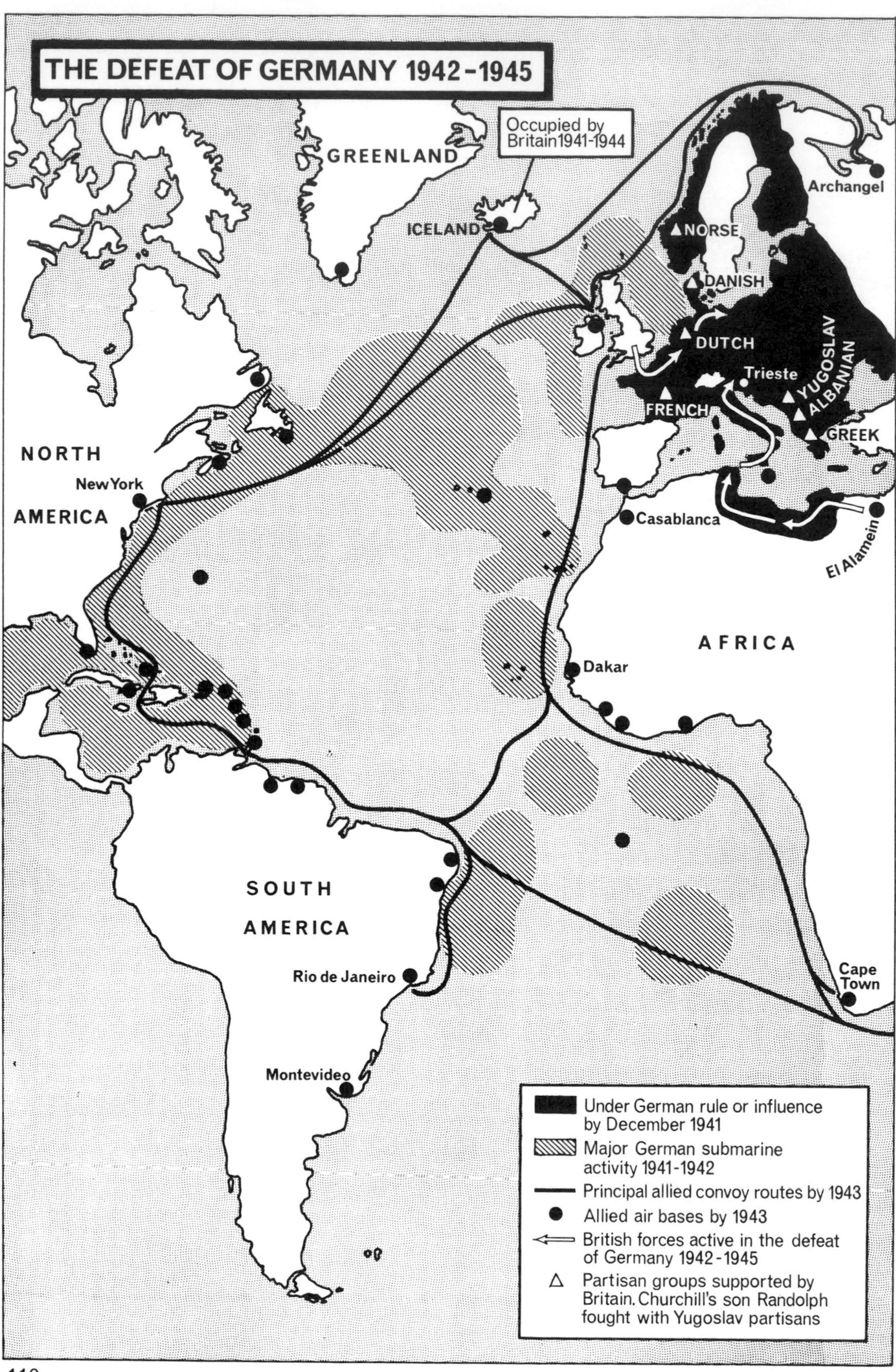
THE DEFEAT OF GERMANY 1942-1945
Occupied by Britain 1941-1944
GREENLAND
ICELAND
Archangel
NORSE
DANISH
DUTCH
YUGOSLAV
ALBANIAN
Trieste
FRENCH
GREEK
NORTH
AMERICA
New York
Casablanca
El Alamein
AFRICA
Dakar
SOUTH
AMERICA
Rio de Janeiro
Cape Town
Montevideo
Under German rule or influence by December 1941
Major German submarine activity 1941-1942
Principal allied convoy routes by 1943
Allied air bases by 1943
British forces active in the defeat of Germany 1942-1945
Partisan groups supported by Britain. Churchill's son Randolph fought with Yugoslav partisans

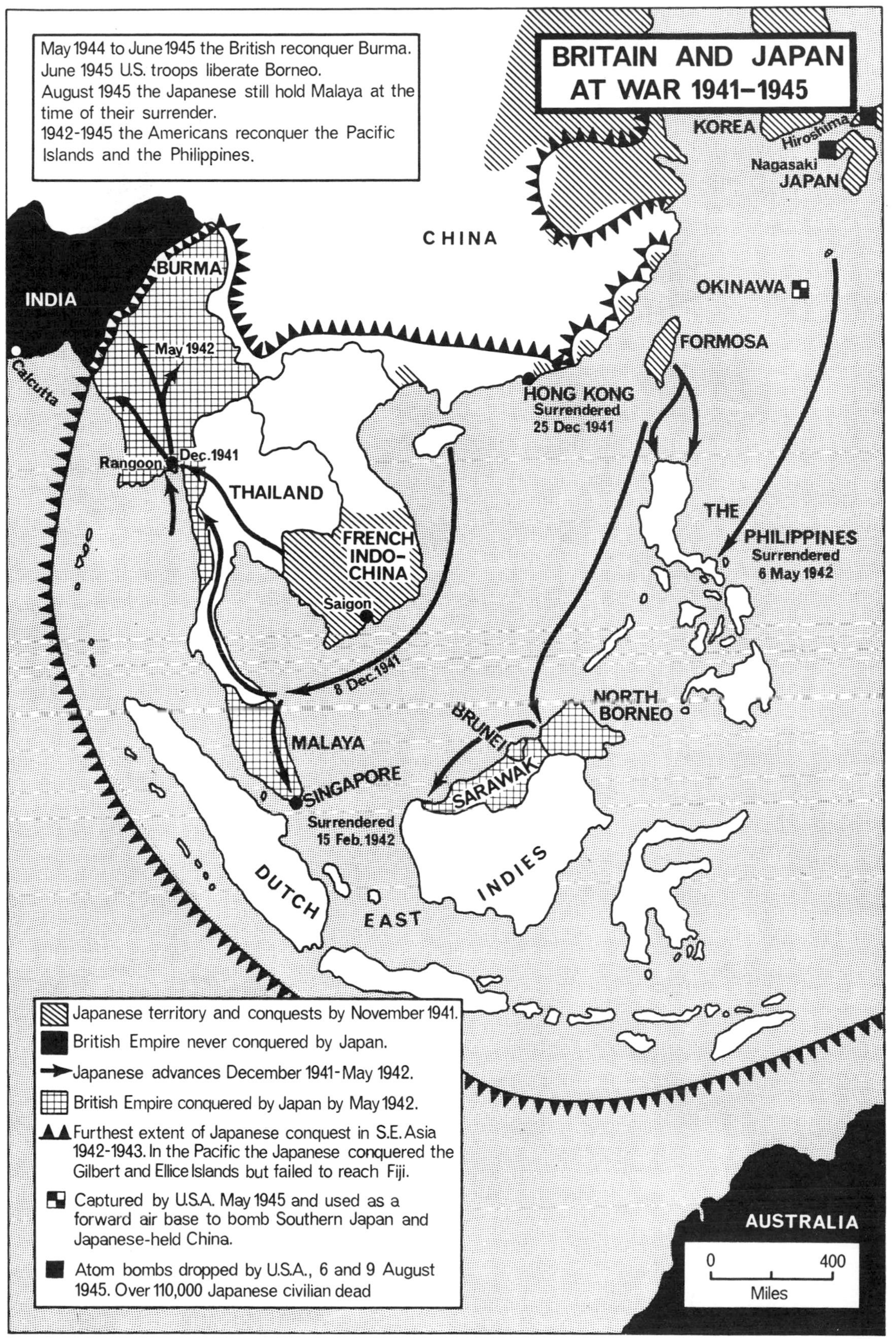

BRITAIN AND JAPAN AT WAR 1941–1945
May 1944 to June 1945 the British reconquer Burma.
June 1945 U.S. troops liberate Borneo.
August 1945 the Japanese still hold Malaya at the time of their surrender.
1942-1945 the Americans reconquer the Pacific Islands and the Philippines.
KOREA
Hiroshima
Nagasaki
JAPAN
CHINA
OKINAWA
FORMOSA
BURMA
INDIA
Calcutta
May 1942
HONG KONG
Surrendered
25 Dec 1941
Rangoon
Dec.1941
THAILAND
THE PHILIPPINES
Surrendered
6 May 1942
FRENCH INDO-CHINA
Saigon
8 Dec 1941
NORTH BORNEO
BRUNEI
MALAYA
SINGAPORE
SARAWAK
Surrendered
15 Feb. 1942
DUTCH
EAST
INDIES
Japanese territory and conquests by November 1941.
British Empire never conquered by Japan.
Japanese advances December 1941-May 1942.
British Empire conquered by Japan by May 1942.
Furthest extent of Japanese conquest in S.E. Asia 1942-1943. In the Pacific the Japanese conquered the Gilbert and Ellice Islands but failed to reach Fiji.
Captured by U.S.A. May 1945 and used as a forward air base to bomb Southern Japan and Japanese-held China.
Atom bombs dropped by U.S.A., 6 and 9 August 1945. Over 110,000 Japanese civilian dead
AUSTRALIA
0
400
Miles

British occupation zones in Germany and Austria 1945 – 48.
European Free Trade Association (EFTA) 1958.
Associate Members of EFTA.
The "Iron Curtain".
European Common Market established by the Treaty of Rome 1957. Britain's first application in 1962 rejected. Second application made in 1967.
Members of the North Atlantic Treaty Organisation (NATO) established 1949. The USA and Canada are also members. Turkey was admitted 1951.
0
400
Miles
FINLAND
February 1947 Anglo–Soviet Peace Treaty limits Army to 34,000 men and Air Force to 60 machines
SWEDEN
NORWAY
U. S. S. R.
DENMARK
EIRE
GREAT BRITAIN
NETHERLANDS
Berlin
POLAND
GERMAN DEMOCRATIC REPUBLIC
GERMAN FEDERAL REPUBLIC
BELGIUM
LUXEMBOURG
CZECHOSLOVAKIA
HUNGARY
AUSTRIA
RUMANIA
FRANCE
SWITZ.
YUGOSLAVIA
BULGARIA
ITALY
ALBANIA
GREECE
PORTUGAL
SPAIN
GIBRALTAR
Anglo–Spanish dispute over sovereignty
BRITAIN AND EUROPE 1945–1965

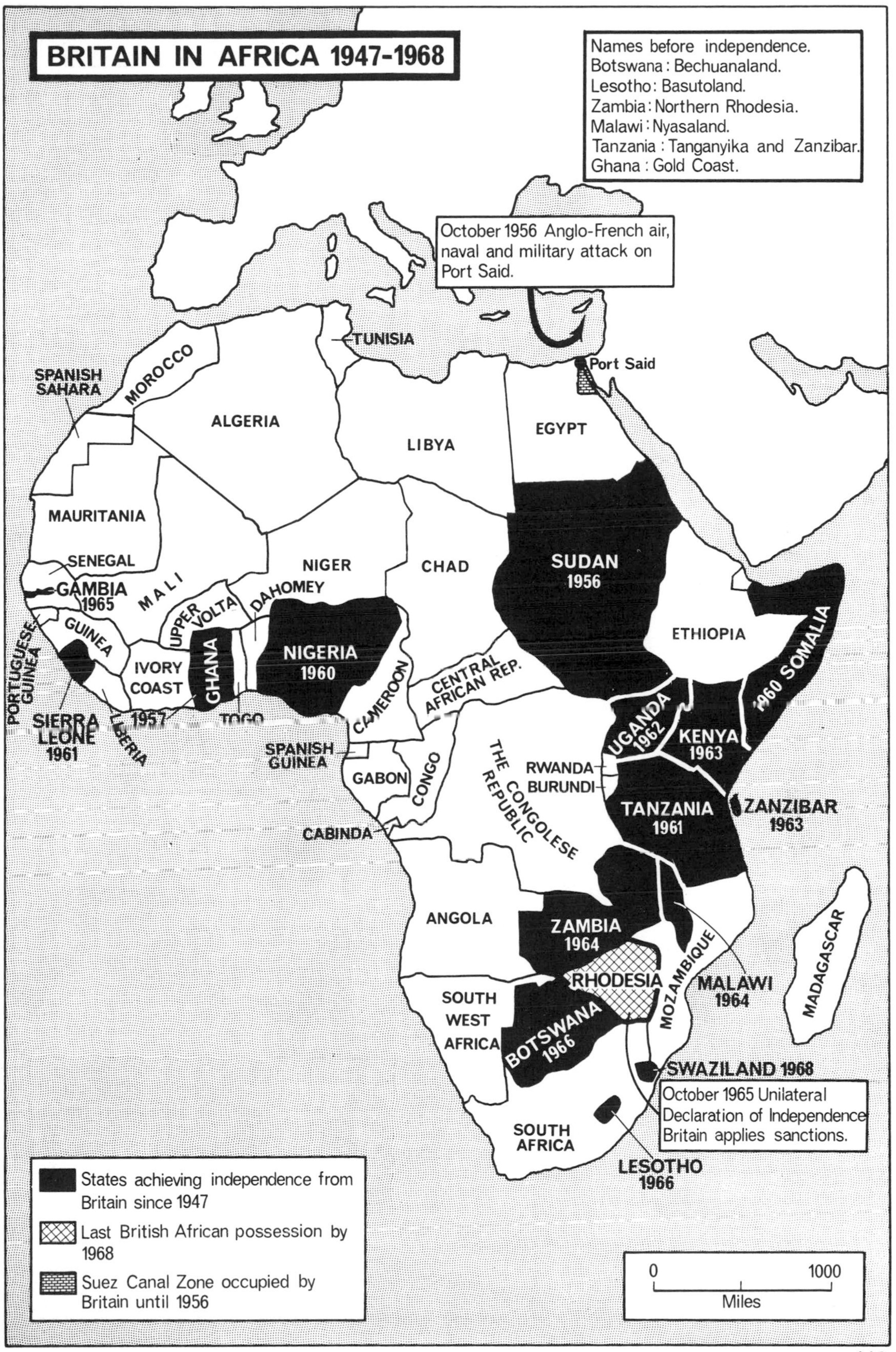
BRITAIN IN AFRICA 1947-1968
Names before independence.
Botswana : Bechuanaland.
Lesotho : Basutoland.
Zambia : Northern Rhodesia.
Malawi : Nyasaland.
Tanzania : Tanganyika and Zanzibar.
Ghana : Gold Coast.
October 1956 Anglo-French air, naval and military attack on Port Said.
Port Said
TUNISIA
MOROCCO
SPANISH SAHARA
ALGERIA
LIBYA
EGYPT
MAURITANIA
SENEGAL
GAMBIA 1965
MALI
NIGER
CHAD
SUDAN 1956
UPPER VOLTA
DAHOMEY
GUINEA
PORTUGUESE GUINEA
IVORY COAST
GHANA 1957
TOGO
NIGERIA 1960
CAMEROON
CENTRAL AFRICAN REP.
ETHIOPIA
SOMALIA 1960
SIERRA LEONE 1961
LIBERIA
SPANISH GUINEA
GABON
CONGO
THE CONGOLESE REPUBLIC
RWANDA
BURUNDI
UGANDA 1962
KENYA 1963
TANZANIA 1961
ZANZIBAR 1963
CABINDA
ANGOLA
ZAMBIA 1964
RHODESIA
MOZAMBIQUE
MALAWI 1964
MADAGASCAR
SOUTH WEST AFRICA
BOTSWANA 1966
SWAZILAND 1968
October 1965 Unilateral Declaration of Independence Britain applies sanctions.
SOUTH AFRICA
LESOTHO 1966
States achieving independence from Britain since 1947
Last British African possession by 1968
Suez Canal Zone occupied by Britain until 1956
0
1000
Miles

UNIVERSITY FOUNDATIONS 1264–1967
0
50
Miles
Aberdeen 1495
Dundee 1967
St. Andrews 1410
1967 Stirling
Glasgow 1451
Strathclyde 1964
Edinburgh 1583
Heriot-Watt 1966
Newcastle 1963
Durham 1832
Lancaster 1964
York 1963
Leeds 1904
Hull 1954
Bradford 1966
Liverpool 1903
Manchester 1851
Salford 1967
Sheffield 1905
Bangor
Keele 1962
Nottingham 1938
1966 Loughborough
Leicester 1957
East Anglia 1964
Aston 1966
Birmingham 1900
University of Wales 1893
Aberystwyth
Warwick 1965
Cambridge 1284
Essex 1965
Oxford 1264
Swansea
Cardiff
Brunel 1966
Reading 1926
Surrey 1966
London 1836
The City University 1966
Bristol 1909
Bath 1966
Kent 1965
Southampton 1952
Sussex 1961
Exeter 1955
Founded 1264–1583
Nineteenth century foundations
Founded 1900–1938
Founded 1952–1967

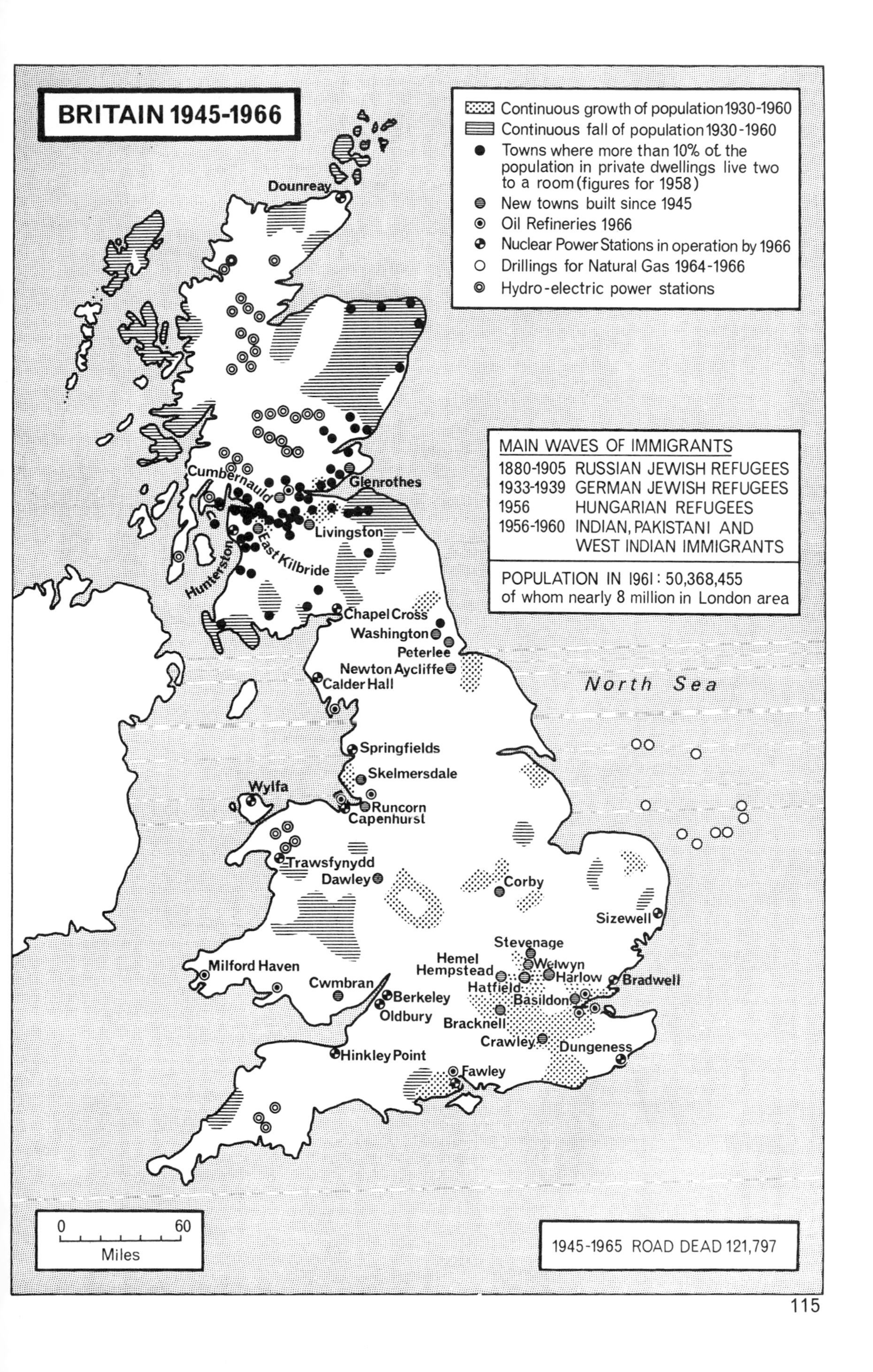

BRITAIN 1945-1966
Continuous growth of population 1930-1960
Continuous fall of population 1930-1960
Towns where more than 10% of the population in private dwellings live two to a room (figures for 1958)
New towns built since 1945
Oil Refineries 1966
Nuclear Power Stations in operation by 1966
Drillings for Natural Gas 1964-1966
Hydro-electric power stations
MAIN WAVES OF IMMIGRANTS
1880-1905 RUSSIAN JEWISH REFUGEES
1933-1939 GERMAN JEWISH REFUGEES
1956 HUNGARIAN REFUGEES
1956-1960 INDIAN, PAKISTANI AND WEST INDIAN IMMIGRANTS
POPULATION IN 1961: 50,368,455
of whom nearly 8 million in London area
North Sea
Dounreay
Cumbernauld
Glenrothes
Livingston
East Kilbride
Hunterston
Chapel Cross
Washington
Peterlee
Newton Aycliffe
Calder Hall
Springfields
Skelmersdale
Wylfa
Runcorn
Capenhurst
Trawsfynydd
Dawley
Corby
Sizewell
Stevenage
Welwyn
Hemel Hempstead
Harlow
Bradwell
Hatfield
Basildon
Milford Haven
Cwmbran
Berkeley
Oldbury
Bracknell
Crawley
Dungeness
Hinkley Point
Fawley
0 60
Miles
1945-1965 ROAD DEAD 121,797

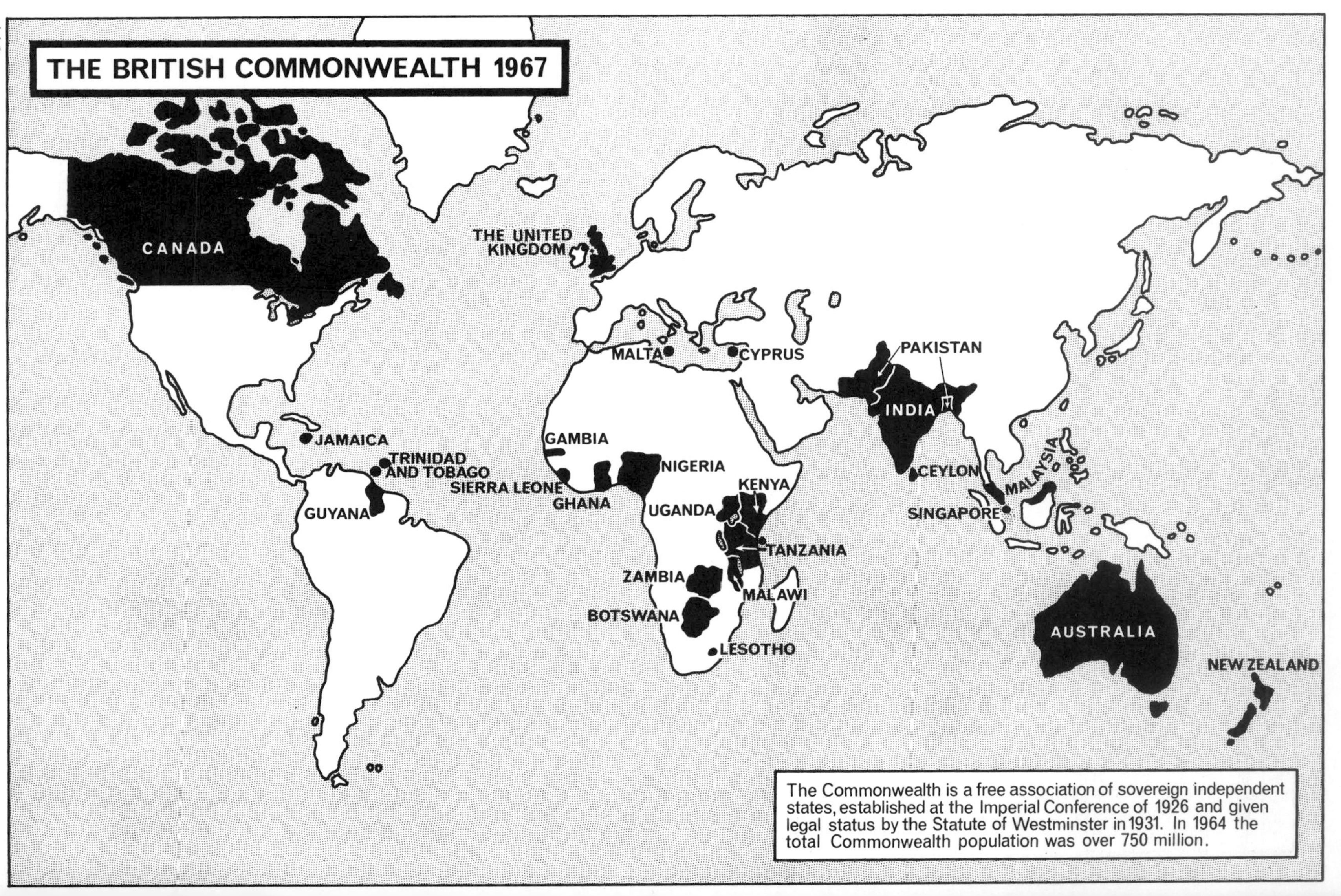
THE BRITISH COMMONWEALTH 1967
CANADA
THE UNITED KINGDOM
MALTA
CYPRUS
PAKISTAN
INDIA
JAMAICA
TRINIDAD AND TOBAGO
GUYANA
GAMBIA
SIERRA LEONE
GHANA
NIGERIA
KENYA
UGANDA
TANZANIA
ZAMBIA
MALAWI
BOTSWANA
LESOTHO
CEYLON
MALAYSIA
SINGAPORE
AUSTRALIA
NEW ZEALAND
The Commonwealth is a free association of sovereign independent states, established at the Imperial Conference of 1926 and given legal status by the Statute of Westminster in 1931. In 1964 the total Commonwealth population was over 750 million.

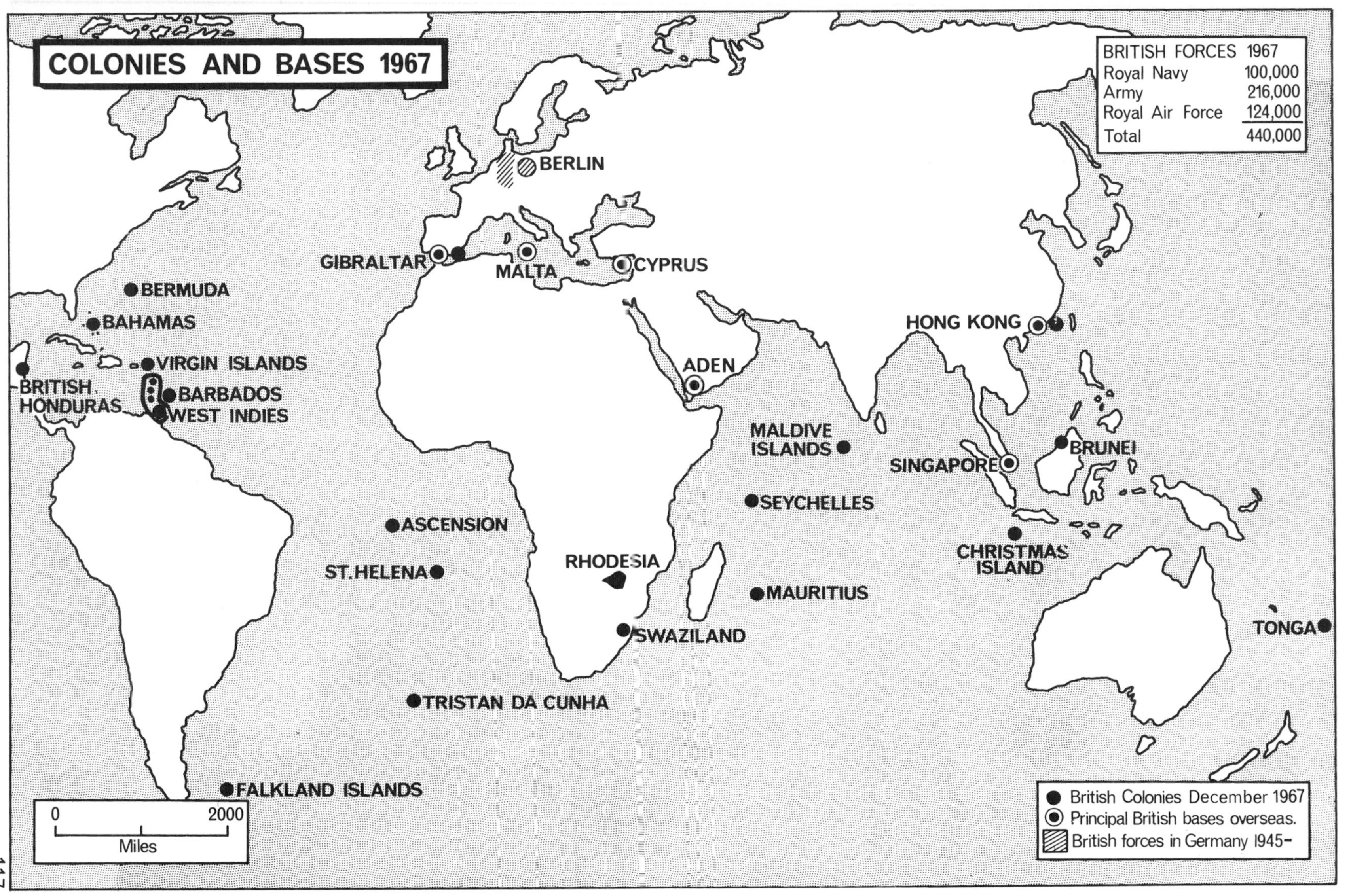
COLONIES AND BASES 1967
BRITISH FORCES 1967
Royal Navy 100,000
Army 216,000
Royal Air Force 124,000
Total 440,000
BERLIN
GIBRALTAR
MALTA
CYPRUS
BERMUDA
BAHAMAS
VIRGIN ISLANDS
BRITISH HONDURAS
BARBADOS
WEST INDIES
ADEN
HONG KONG
MALDIVE ISLANDS
SINGAPORE
BRUNEI
SEYCHELLES
ASCENSION
ST. HELENA
RHODESIA
CHRISTMAS ISLAND
MAURITIUS
SWAZILAND
TONGA
TRISTAN DA CUNHA
FALKLAND ISLANDS
0
2000
Miles
British Colonies December 1967
Principal British bases overseas.
British forces in Germany 1945–

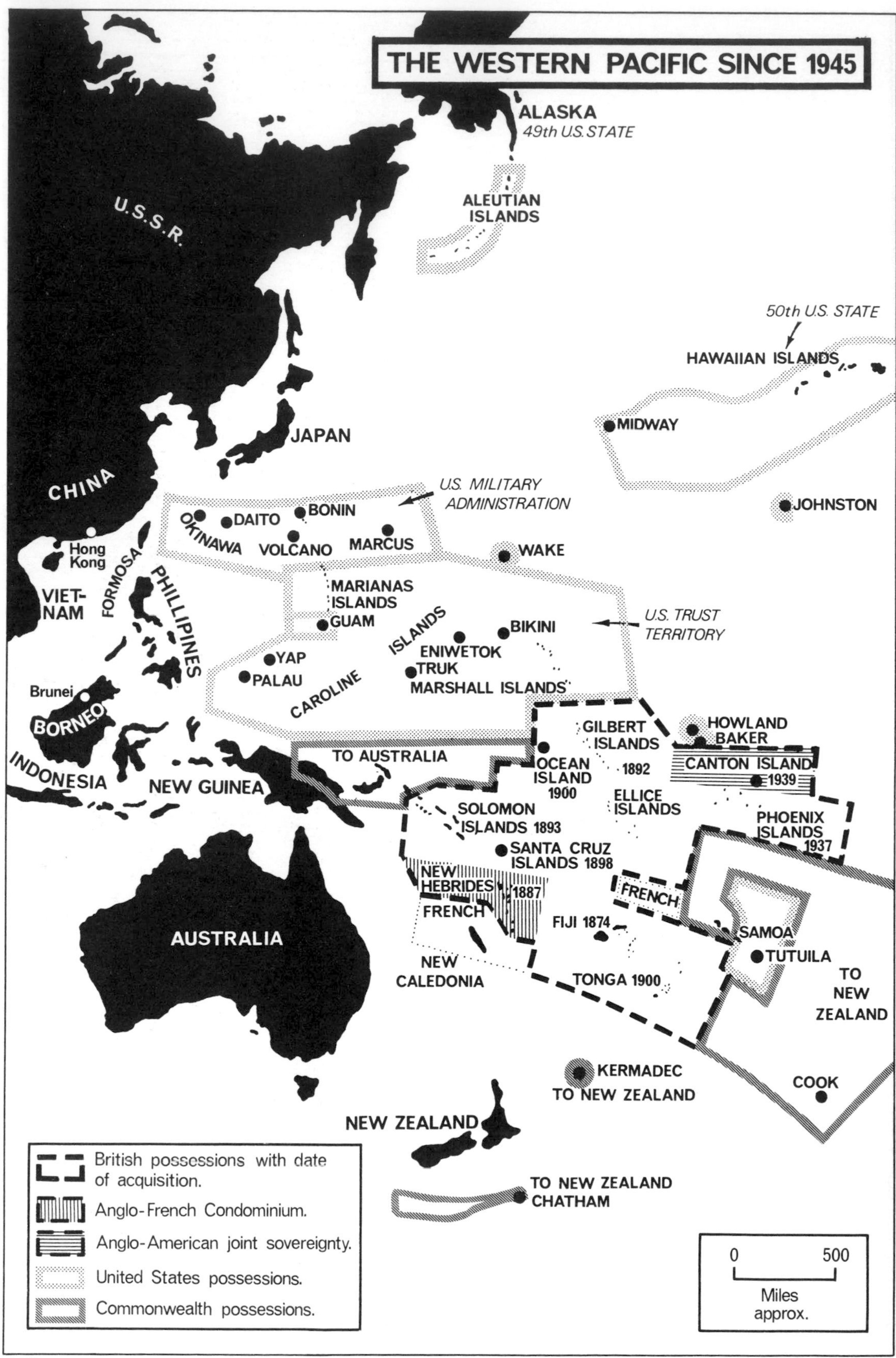
THE WESTERN PACIFIC SINCE 1945
ALASKA
49th U.S. STATE
ALEUTIAN ISLANDS
U.S.S.R.
50th U.S. STATE
HAWAIIAN ISLANDS
MIDWAY
JAPAN
CHINA
U.S. MILITARY ADMINISTRATION
JOHNSTON
BONIN
DAITO
OKINAWA
VOLCANO
MARCUS
WAKE
Hong Kong
FORMOSA
PHILIPPINES
VIET-NAM
MARIANAS ISLANDS
GUAM
U.S. TRUST TERRITORY
CAROLINE ISLANDS
BIKINI
ENIWETOK
YAP
TRUK
PALAU
MARSHALL ISLANDS
Brunei
BORNEO
GILBERT ISLANDS
HOWLAND
BAKER
TO AUSTRALIA
OCEAN ISLAND 1900
1892
CANTON ISLAND
1939
INDONESIA
NEW GUINEA
ELLICE ISLANDS
SOLOMON ISLANDS 1893
PHOENIX ISLANDS 1937
SANTA CRUZ ISLANDS 1898
NEW HEBRIDES
1887
FRENCH
FRENCH
SAMOA
AUSTRALIA
FIJI 1874
TUTUILA
NEW CALEDONIA
TONGA 1900
TO NEW ZEALAND
KERMADEC
TO NEW ZEALAND
COOK
NEW ZEALAND
British possessions with date of acquisition.
Anglo-French Condominium.
Anglo-American joint sovereignty.
United States possessions.
Commonwealth possessions.
TO NEW ZEALAND
CHATHAM
0
500
Miles approx.

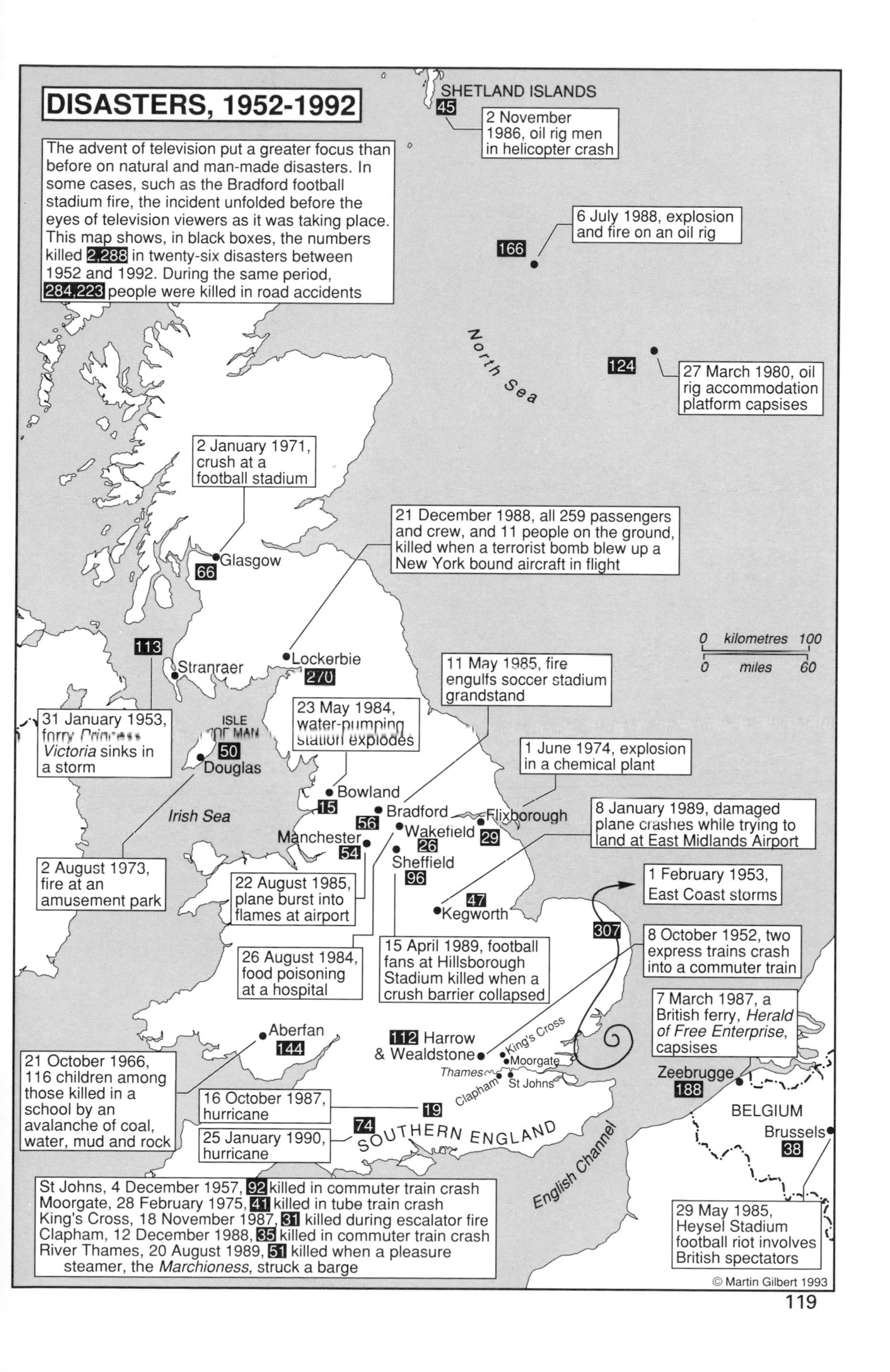

DISASTERS, 1952-1992
The advent of television put a greater focus than before on natural and man-made disasters. In some cases, such as the Bradford football stadium fire, the incident unfolded before the eyes of television viewers as it was taking place. This map shows, in black boxes, the numbers killed 2,288 in twenty-six disasters between 1952 and 1992. During the same period, 284,223 people were killed in road accidents
SHETLAND ISLANDS
45
2 November 1986, oil rig men in helicopter crash
6 July 1988, explosion and fire on an oil rig
166
North Sea
124
27 March 1980, oil rig accommodation platform capsises
2 January 1971, crush at a football stadium
Glasgow
66
21 December 1988, all 259 passengers and crew, and 11 people on the ground, killed when a terrorist bomb blew up a New York bound aircraft in flight
113
Stranraer
Lockerbie
270
0 kilometres 100
0 miles 60
11 May 1985, fire engulfs soccer stadium grandstand
31 January 1953,
Victoria sinks in a storm
23 May 1984, water-pumping station explodes
50
Douglas
1 June 1974, explosion in a chemical plant
Bowland
15
Irish Sea
Bradford
56
Flixborough
29
8 January 1989, damaged plane crashes while trying to land at East Midlands Airport
Manchester
54
Wakefield
26
Sheffield
96
2 August 1973, fire at an amusement park
22 August 1985, plane burst into flames at airport
47
Kegworth
1 February 1953, East Coast storms
307
8 October 1952, two express trains crash into a commuter train
26 August 1984, food poisoning at a hospital
15 April 1989, football fans at Hillsborough Stadium killed when a crush barrier collapsed
7 March 1987, a British ferry, Herald of Free Enterprise, capsises
Aberfan
144
112 Harrow & Wealdstone
King's Cross
Moorgate
Thames
Clapham
St Johns
Zeebrugge
188
21 October 1966, 116 children among those killed in a school by an avalanche of coal, water, mud and rock
16 October 1987, hurricane
19
BELGIUM
Brussels
38
74
25 January 1990, hurricane
SOUTHERN ENGLAND
English Channel
St Johns, 4 December 1957, 92 killed in commuter train crash
Moorgate, 28 February 1975, 41 killed in tube train crash
King's Cross, 18 November 1987, 31 killed during escalator fire
Clapham, 12 December 1988, 35 killed in commuter train crash
River Thames, 20 August 1989, 51 killed when a pleasure steamer, the Marchioness, struck a barge
29 May 1985, Heysel Stadium football riot involves British spectators
© Martin Gilbert 1993

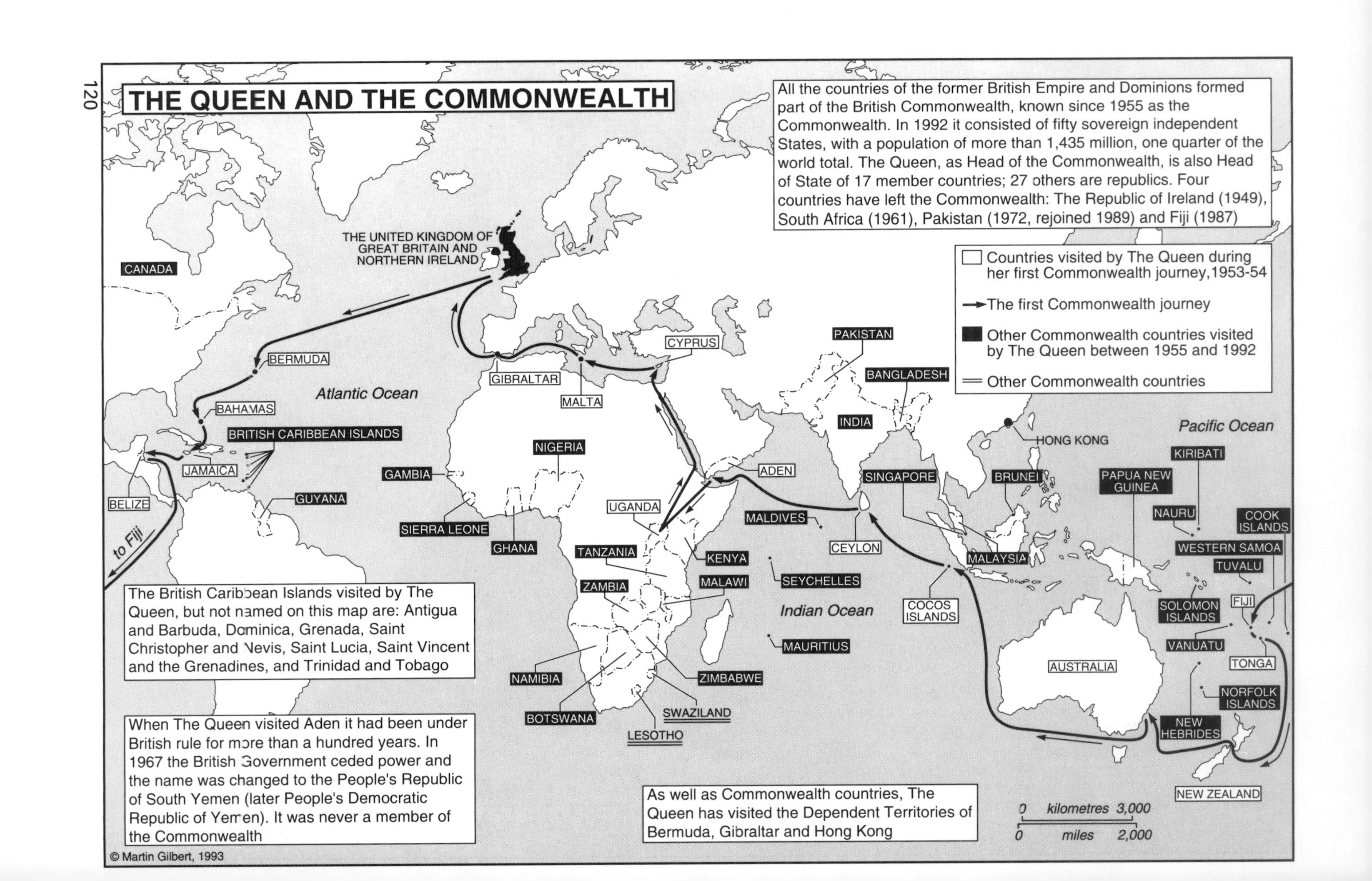
THE QUEEN AND THE COMMONWEALTH
All the countries of the former British Empire and Dominions formed part of the British Commonwealth, known since 1955 as the Commonwealth. In 1992 it consisted of fifty sovereign independent States, with a population of more than 1,435 million, one quarter of the world total. The Queen, as Head of the Commonwealth, is also Head of State of 17 member countries; 27 others are republics. Four countries have left the Commonwealth: The Republic of Ireland (1949), South Africa (1961), Pakistan (1972, rejoined 1989) and Fiji (1987)
Countries visited by The Queen during her first Commonwealth journey, 1953-54
The first Commonwealth journey
Other Commonwealth countries visited by The Queen between 1955 and 1992
Other Commonwealth countries
THE UNITED KINGDOM OF GREAT BRITAIN AND NORTHERN IRELAND
CANADA
BERMUDA
GIBRALTAR
MALTA
CYPRUS
PAKISTAN
BANGLADESH
INDIA
Atlantic Ocean
BAHAMAS
BRITISH CARIBBEAN ISLANDS
JAMAICA
BELIZE
GUYANA
to Fiji
NIGERIA
GAMBIA
SIERRA LEONE
GHANA
UGANDA
ADEN
TANZANIA
KENYA
ZAMBIA
MALAWI
MALDIVES
CEYLON
SEYCHELLES
MAURITIUS
Indian Ocean
NAMIBIA
ZIMBABWE
BOTSWANA
SWAZILAND
LESOTHO
SINGAPORE
MALAYSIA
BRUNEI
HONG KONG
COCOS ISLANDS
Pacific Ocean
KIRIBATI
PAPUA NEW GUINEA
NAURU
COOK ISLANDS
WESTERN SAMOA
TUVALU
SOLOMON ISLANDS
FIJI
VANUATU
TONGA
AUSTRALIA
NORFOLK ISLANDS
NEW HEBRIDES
NEW ZEALAND
The British Caribbean Islands visited by The Queen, but not named on this map are: Antigua and Barbuda, Dominica, Grenada, Saint Christopher and Nevis, Saint Lucia, Saint Vincent and the Grenadines, and Trinidad and Tobago
When The Queen visited Aden it had been under British rule for more than a hundred years. In 1967 the British Government ceded power and the name was changed to the People's Republic of South Yemen (later People's Democratic Republic of Yemen). It was never a member of the Commonwealth
As well as Commonwealth countries, The Queen has visited the Dependent Territories of Bermuda, Gibraltar and Hong Kong
0 kilometres 3,000
0 miles 2,000
© Martin Gilbert, 1993

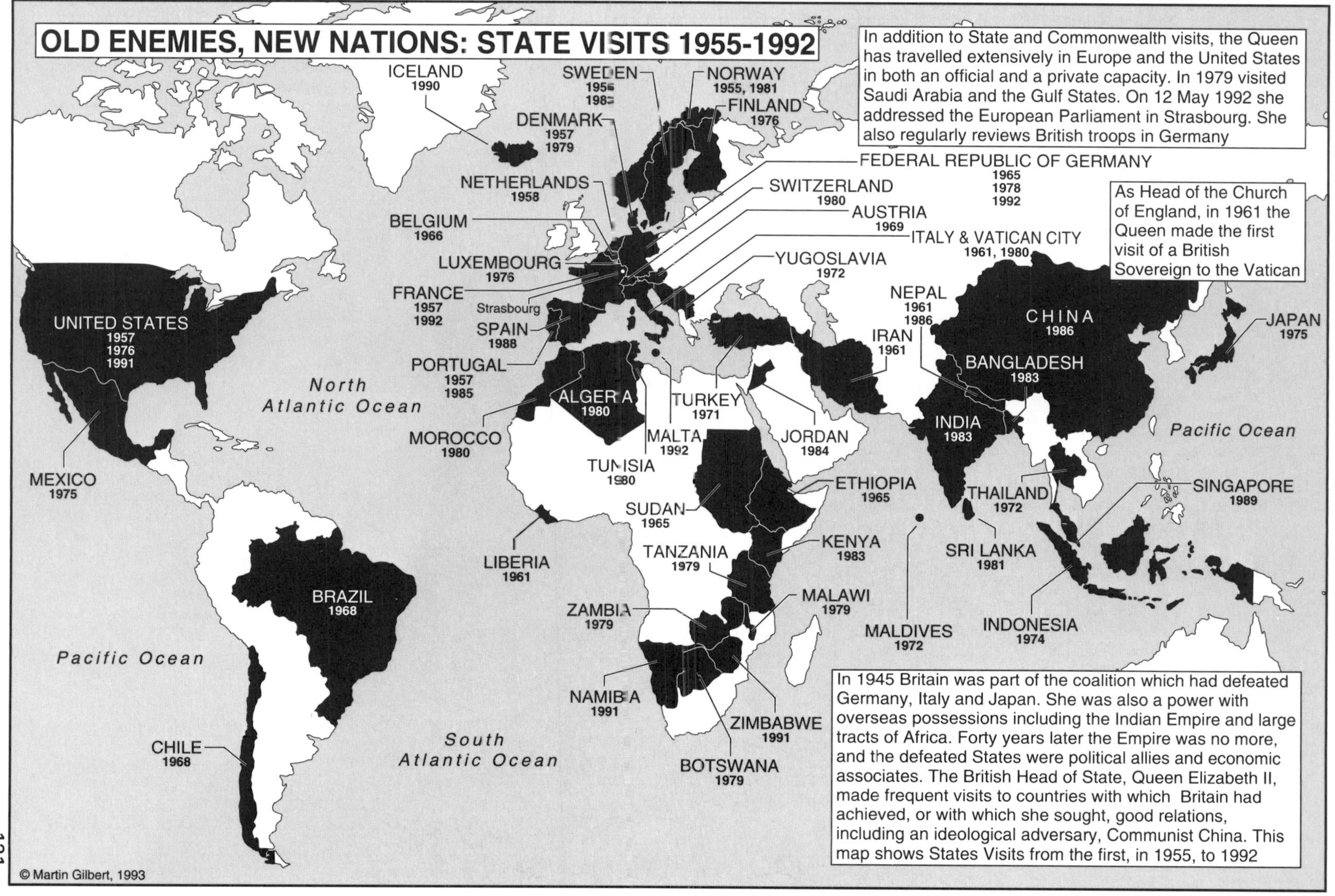

OLD ENEMIES, NEW NATIONS: STATE VISITS 1955-1992
In addition to State and Commonwealth visits, the Queen has travelled extensively in Europe and the United States in both an official and a private capacity. In 1979 visited Saudi Arabia and the Gulf States. On 12 May 1992 she addressed the European Parliament in Strasbourg. She also regularly reviews British troops in Germany
As Head of the Church of England, in 1961 the Queen made the first visit of a British Sovereign to the Vatican
In 1945 Britain was part of the coalition which had defeated Germany, Italy and Japan. She was also a power with overseas possessions including the Indian Empire and large tracts of Africa. Forty years later the Empire was no more, and the defeated States were political allies and economic associates. The British Head of State, Queen Elizabeth II, made frequent visits to countries with which Britain had achieved, or with which she sought, good relations, including an ideological adversary, Communist China. This map shows States Visits from the first, in 1955, to 1992
ICELAND 1990
SWEDEN
NORWAY 1955, 1981
FINLAND 1976
DENMARK 1957 1979
NETHERLANDS 1958
BELGIUM 1966
LUXEMBOURG 1976
FRANCE 1957 1992
Strasbourg
SPAIN 1988
PORTUGAL 1957 1985
MOROCCO 1980
FEDERAL REPUBLIC OF GERMANY 1965 1978 1992
SWITZERLAND 1980
AUSTRIA 1969
ITALY & VATICAN CITY 1961, 1980
YUGOSLAVIA 1972
TURKEY 1971
MALTA 1992
TUNISIA 1980
ALGERIA 1980
JORDAN 1984
IRAN 1961
NEPAL 1961 1986
CHINA 1986
BANGLADESH 1983
INDIA 1983
JAPAN 1975
Pacific Ocean
SINGAPORE 1989
THAILAND 1972
SRI LANKA 1981
INDONESIA 1974
MALDIVES 1972
ETHIOPIA 1965
SUDAN 1965
KENYA 1983
TANZANIA 1979
MALAWI 1979
ZAMBIA 1979
NAMIBIA 1991
ZIMBABWE 1991
BOTSWANA 1979
LIBERIA 1961
UNITED STATES 1957 1976 1991
MEXICO 1975
North Atlantic Ocean
BRAZIL 1968
CHILE 1968
Pacific Ocean
South Atlantic Ocean
© Martin Gilbert, 1993

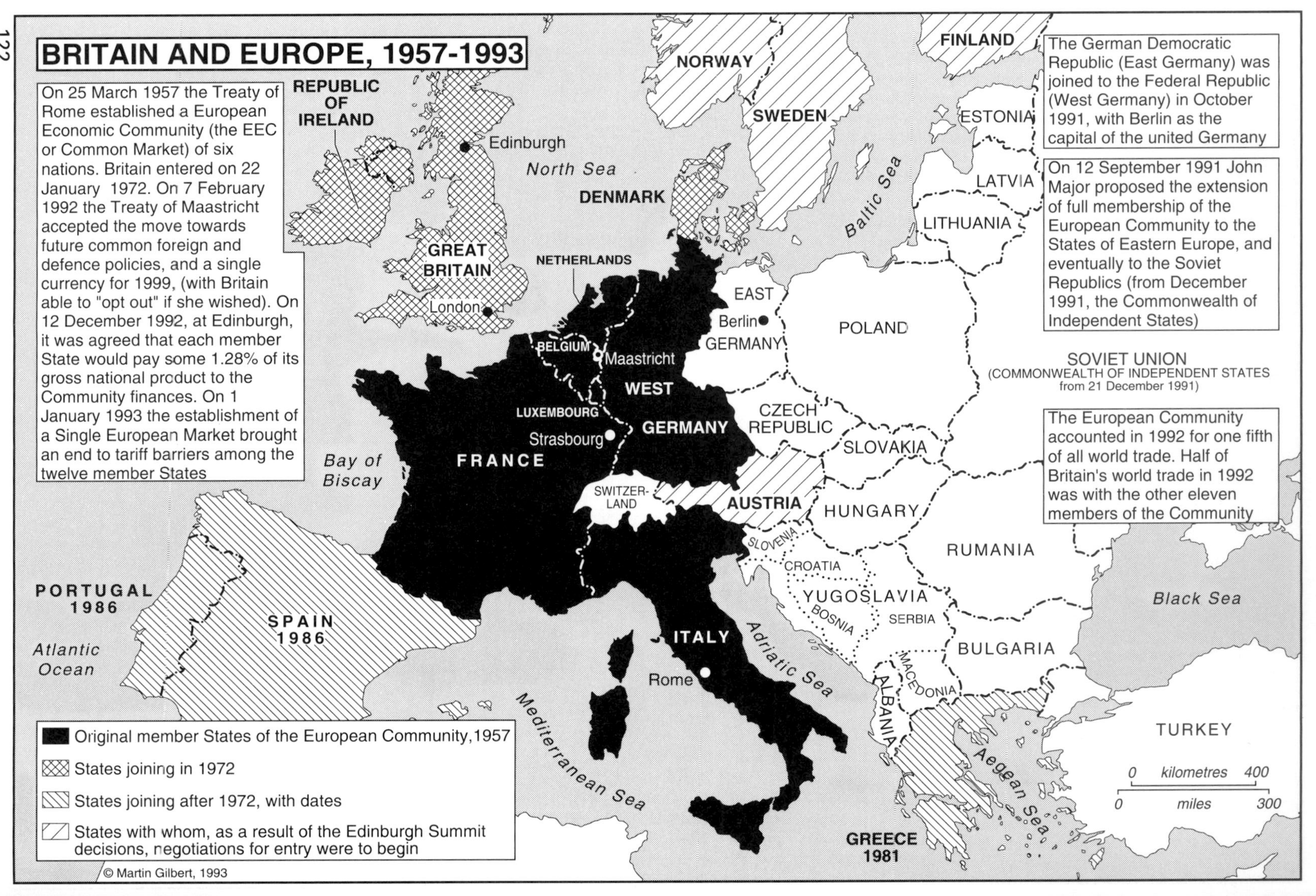

BRITAIN AND EUROPE, 1957-1993
On 25 March 1957 the Treaty of Rome established a European Economic Community (the EEC or Common Market) of six nations. Britain entered on 22 January 1972. On 7 February 1992 the Treaty of Maastricht accepted the move towards future common foreign and defence policies, and a single currency for 1999, (with Britain able to "opt out" if she wished). On 12 December 1992, at Edinburgh, it was agreed that each member State would pay some 1.28% of its gross national product to the Community finances. On 1 January 1993 the establishment of a Single European Market brought an end to tariff barriers among the twelve member States
The German Democratic Republic (East Germany) was joined to the Federal Republic (West Germany) in October 1991, with Berlin as the capital of the united Germany
On 12 September 1991 John Major proposed the extension of full membership of the European Community to the States of Eastern Europe, and eventually to the Soviet Republics (from December 1991, the Commonwealth of Independent States)
The European Community accounted in 1992 for one fifth of all world trade. Half of Britain's world trade in 1992 was with the other eleven members of the Community
REPUBLIC OF IRELAND
GREAT BRITAIN
Edinburgh
London
North Sea
DENMARK
NORWAY
SWEDEN
FINLAND
ESTONIA
LATVIA
LITHUANIA
Baltic Sea
NETHERLANDS
BELGIUM
Maastricht
LUXEMBOURG
WEST GERMANY
EAST GERMANY
Berlin
POLAND
SOVIET UNION
(COMMONWEALTH OF INDEPENDENT STATES from 21 December 1991)
CZECH REPUBLIC
SLOVAKIA
HUNGARY
RUMANIA
AUSTRIA
SWITZER-LAND
Strasbourg
FRANCE
Bay of Biscay
SLOVENIA
CROATIA
YUGOSLAVIA
BOSNIA
SERBIA
BULGARIA
MACEDONIA
ALBANIA
Black Sea
TURKEY
PORTUGAL 1986
SPAIN 1986
Atlantic Ocean
ITALY
Rome
Adriatic Sea
Mediterranean Sea
Aegean Sea
GREECE 1981
0 kilometres 400
0 miles 300
Original member States of the European Community, 1957
States joining in 1972
States joining after 1972, with dates
States with whom, as a result of the Edinburgh Summit decisions, negotiations for entry were to begin

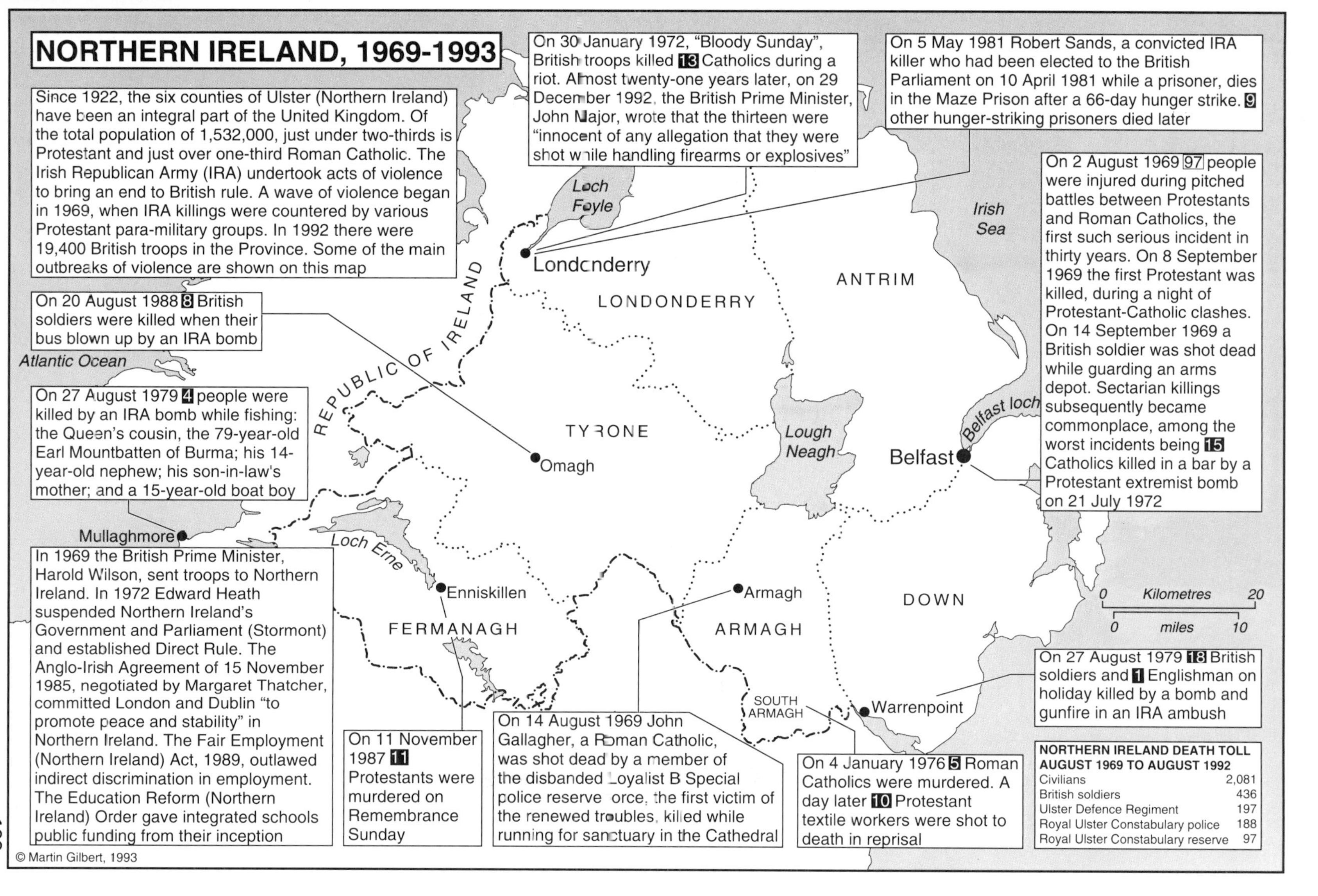

NORTHERN IRELAND, 1969-1993
Since 1922, the six counties of Ulster (Northern Ireland) have been an integral part of the United Kingdom. Of the total population of 1,532,000, just under two-thirds is Protestant and just over one-third Roman Catholic. The Irish Republican Army (IRA) undertook acts of violence to bring an end to British rule. A wave of violence began in 1969, when IRA killings were countered by various Protestant para-military groups. In 1992 there were 19,400 British troops in the Province. Some of the main outbreaks of violence are shown on this map
On 30 January 1972, "Bloody Sunday", British troops killed 13 Catholics during a riot. Almost twenty-one years later, on 29 December 1992, the British Prime Minister, John Major, wrote that the thirteen were "innocent of any allegation that they were shot while handling firearms or explosives"
On 5 May 1981 Robert Sands, a convicted IRA killer who had been elected to the British Parliament on 10 April 1981 while a prisoner, dies in the Maze Prison after a 66-day hunger strike. 9 other hunger-striking prisoners died later
On 2 August 1969 97 people were injured during pitched battles between Protestants and Roman Catholics, the first such serious incident in thirty years. On 8 September 1969 the first Protestant was killed, during a night of Protestant-Catholic clashes. On 14 September 1969 a British soldier was shot dead while guarding an arms depot. Sectarian killings subsequently became commonplace, among the worst incidents being 15 Catholics killed in a bar by a Protestant extremist bomb on 21 July 1972
On 20 August 1988 8 British soldiers were killed when their bus blown up by an IRA bomb
On 27 August 1979 4 people were killed by an IRA bomb while fishing: the Queen's cousin, the 79-year-old Earl Mountbatten of Burma; his 14-year-old nephew; his son-in-law's mother; and a 15-year-old boat boy
In 1969 the British Prime Minister, Harold Wilson, sent troops to Northern Ireland. In 1972 Edward Heath suspended Northern Ireland's Government and Parliament (Stormont) and established Direct Rule. The Anglo-Irish Agreement of 15 November 1985, negotiated by Margaret Thatcher, committed London and Dublin "to promote peace and stability" in Northern Ireland. The Fair Employment (Northern Ireland) Act, 1989, outlawed indirect discrimination in employment. The Education Reform (Northern Ireland) Order gave integrated schools public funding from their inception
On 11 November 1987 11 Protestants were murdered on Remembrance Sunday
On 14 August 1969 John Gallagher, a Roman Catholic, was shot dead by a member of the disbanded Loyalist B Special police reserve force, the first victim of the renewed troubles, killed while running for sanctuary in the Cathedral
On 4 January 1976 5 Roman Catholics were murdered. A day later 10 Protestant textile workers were shot to death in reprisal
On 27 August 1979 18 British soldiers and 1 Englishman on holiday killed by a bomb and gunfire in an IRA ambush
NORTHERN IRELAND DEATH TOLL AUGUST 1969 TO AUGUST 1992
Civilians 2,081
British soldiers 436
Ulster Defence Regiment 197
Royal Ulster Constabulary police 188
Royal Ulster Constabulary reserve 97
Loch Foyle
Londonderry
LONDONDERRY
ANTRIM
Irish Sea
REPUBLIC OF IRELAND
Atlantic Ocean
TYRONE
Omagh
Lough Neagh
Belfast
Belfast loch
Mullaghmore
Loch Erne
Enniskillen
FERMANAGH
Armagh
ARMAGH
DOWN
SOUTH ARMAGH
Warrenpoint
0 Kilometres 20
0 miles 10
© Martin Gilbert, 1993

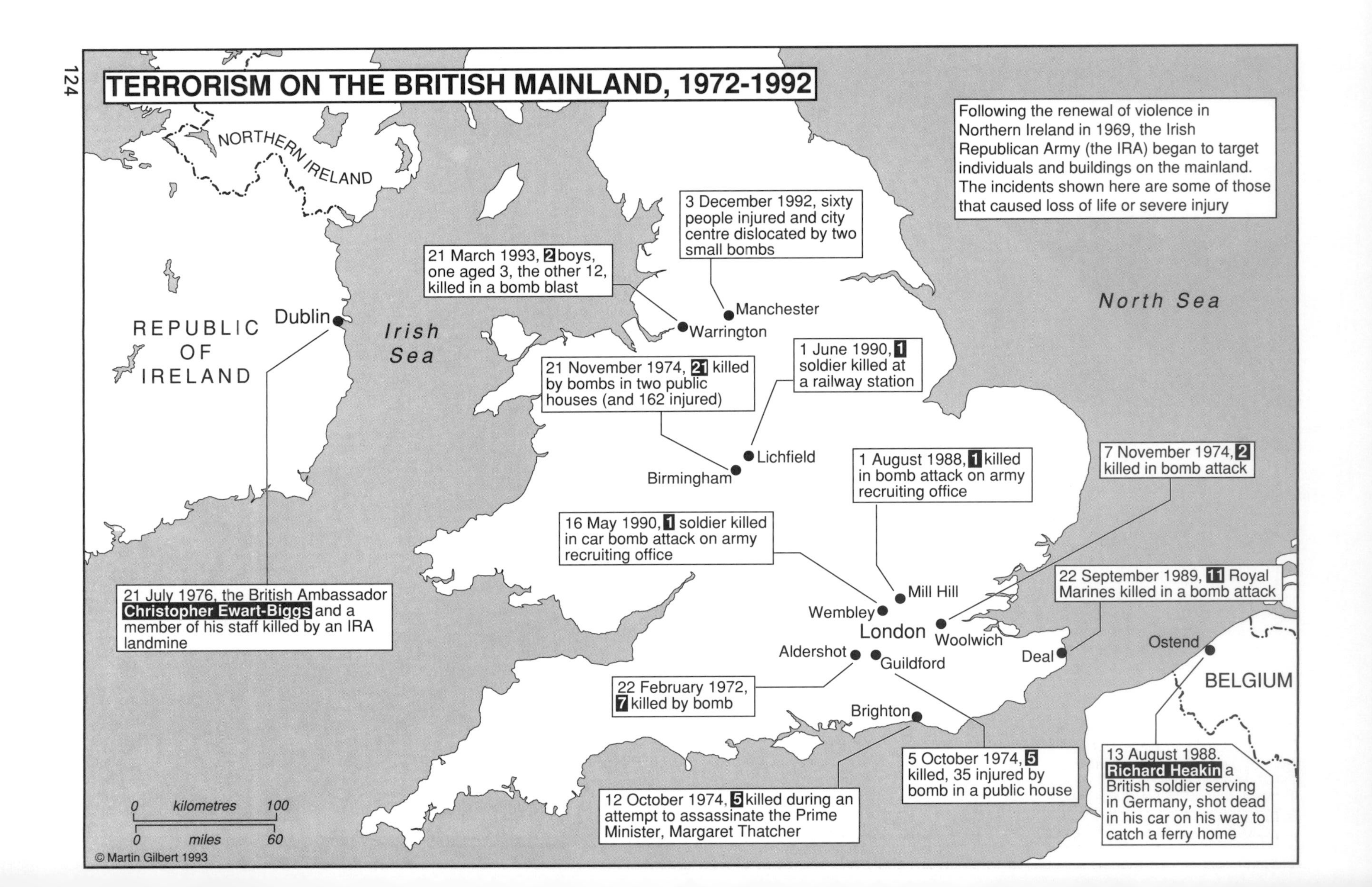
TERRORISM ON THE BRITISH MAINLAND, 1972-1992
Following the renewal of violence in Northern Ireland in 1969, the Irish Republican Army (the IRA) began to target individuals and buildings on the mainland. The incidents shown here are some of those that caused loss of life or severe injury
NORTHERN IRELAND
REPUBLIC OF IRELAND
Dublin
Irish Sea
North Sea
21 March 1993, 2 boys, one aged 3, the other 12, killed in a bomb blast
3 December 1992, sixty people injured and city centre dislocated by two small bombs
Manchester
Warrington
1 June 1990, 1 soldier killed at a railway station
21 November 1974, 21 killed by bombs in two public houses (and 162 injured)
Lichfield
Birmingham
1 August 1988, 1 killed in bomb attack on army recruiting office
7 November 1974, 2 killed in bomb attack
16 May 1990, 1 soldier killed in car bomb attack on army recruiting office
22 September 1989, 11 Royal Marines killed in a bomb attack
Mill Hill
Wembley
London
Woolwich
Aldershot
Guildford
Deal
Ostend
BELGIUM
21 July 1976, the British Ambassador Christopher Ewart-Biggs and a member of his staff killed by an IRA landmine
22 February 1972, 7 killed by bomb
Brighton
5 October 1974, 5 killed, 35 injured by bomb in a public house
13 August 1988. Richard Heakin a British soldier serving in Germany, shot dead in his car on his way to catch a ferry home
12 October 1974, 5 killed during an attempt to assassinate the Prime Minister, Margaret Thatcher
0 kilometres 100
0 miles 60

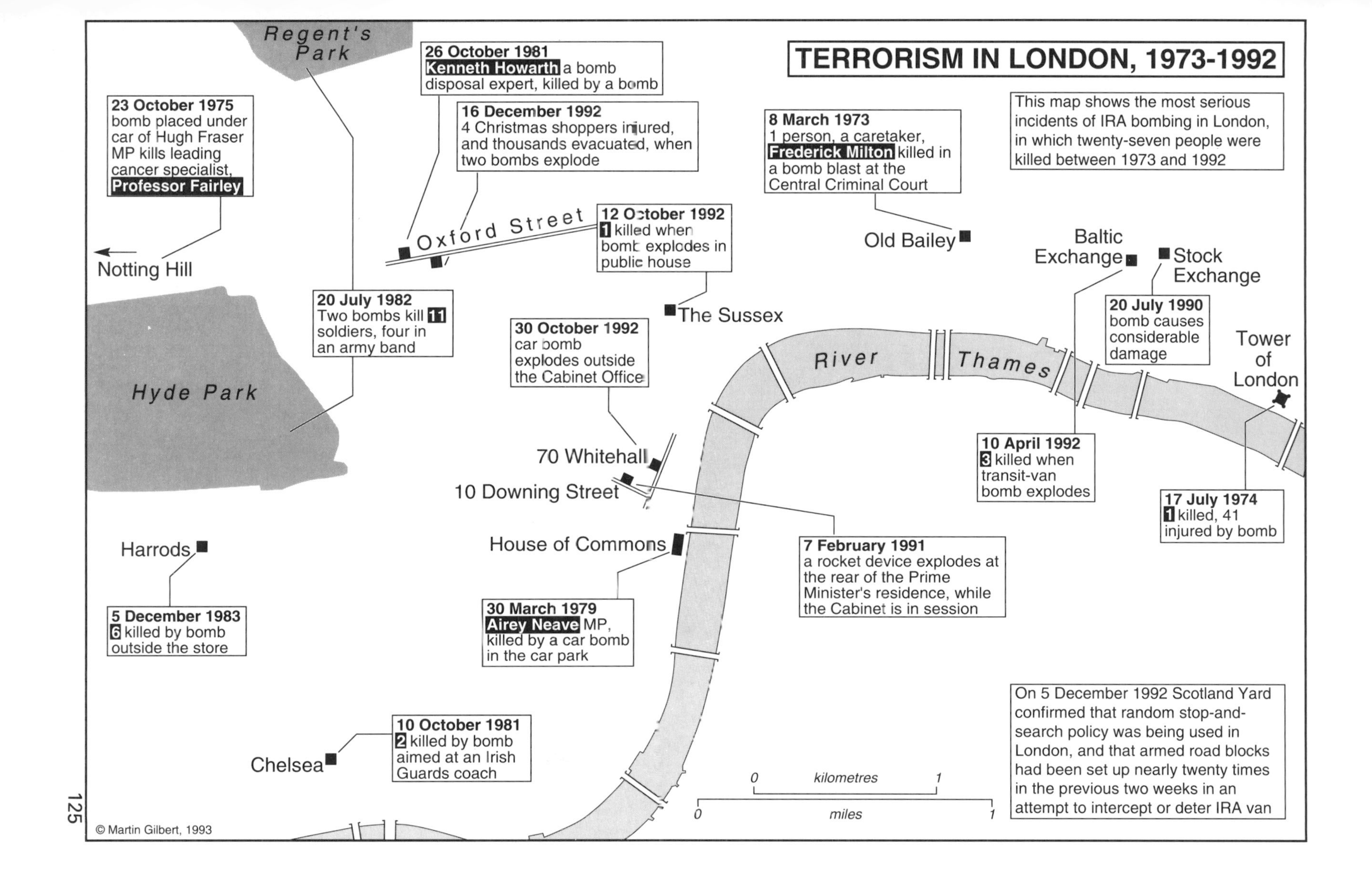
TERRORISM IN LONDON, 1973-1992
This map shows the most serious incidents of IRA bombing in London, in which twenty-seven people were killed between 1973 and 1992
Regent's Park
26 October 1981 Kenneth Howarth a bomb disposal expert, killed by a bomb
23 October 1975 bomb placed under car of Hugh Fraser MP kills leading cancer specialist, Professor Fairley
16 December 1992 4 Christmas shoppers injured, and thousands evacuated, when two bombs explode
8 March 1973 1 person, a caretaker, Frederick Milton killed in a bomb blast at the Central Criminal Court
Oxford Street
12 October 1992 1 killed when bomb explodes in public house
Old Bailey
Baltic Exchange
Stock Exchange
Notting Hill
20 July 1982 Two bombs kill 11 soldiers, four in an army band
The Sussex
20 July 1990 bomb causes considerable damage
Tower of London
30 October 1992 car bomb explodes outside the Cabinet Office
Hyde Park
River Thames
70 Whitehall
10 Downing Street
10 April 1992 3 killed when transit-van bomb explodes
17 July 1974 1 killed, 41 injured by bomb
Harrods
House of Commons
7 February 1991 a rocket device explodes at the rear of the Prime Minister's residence, while the Cabinet is in session
5 December 1983 6 killed by bomb outside the store
30 March 1979 Airey Neave MP, killed by a car bomb in the car park
On 5 December 1992 Scotland Yard confirmed that random stop-and-search policy was being used in London, and that armed road blocks had been set up nearly twenty times in the previous two weeks in an attempt to intercept or deter IRA van
10 October 1981 2 killed by bomb aimed at an Irish Guards coach
Chelsea
0 kilometres 1
0 miles 1
© Martin Gilbert, 1993

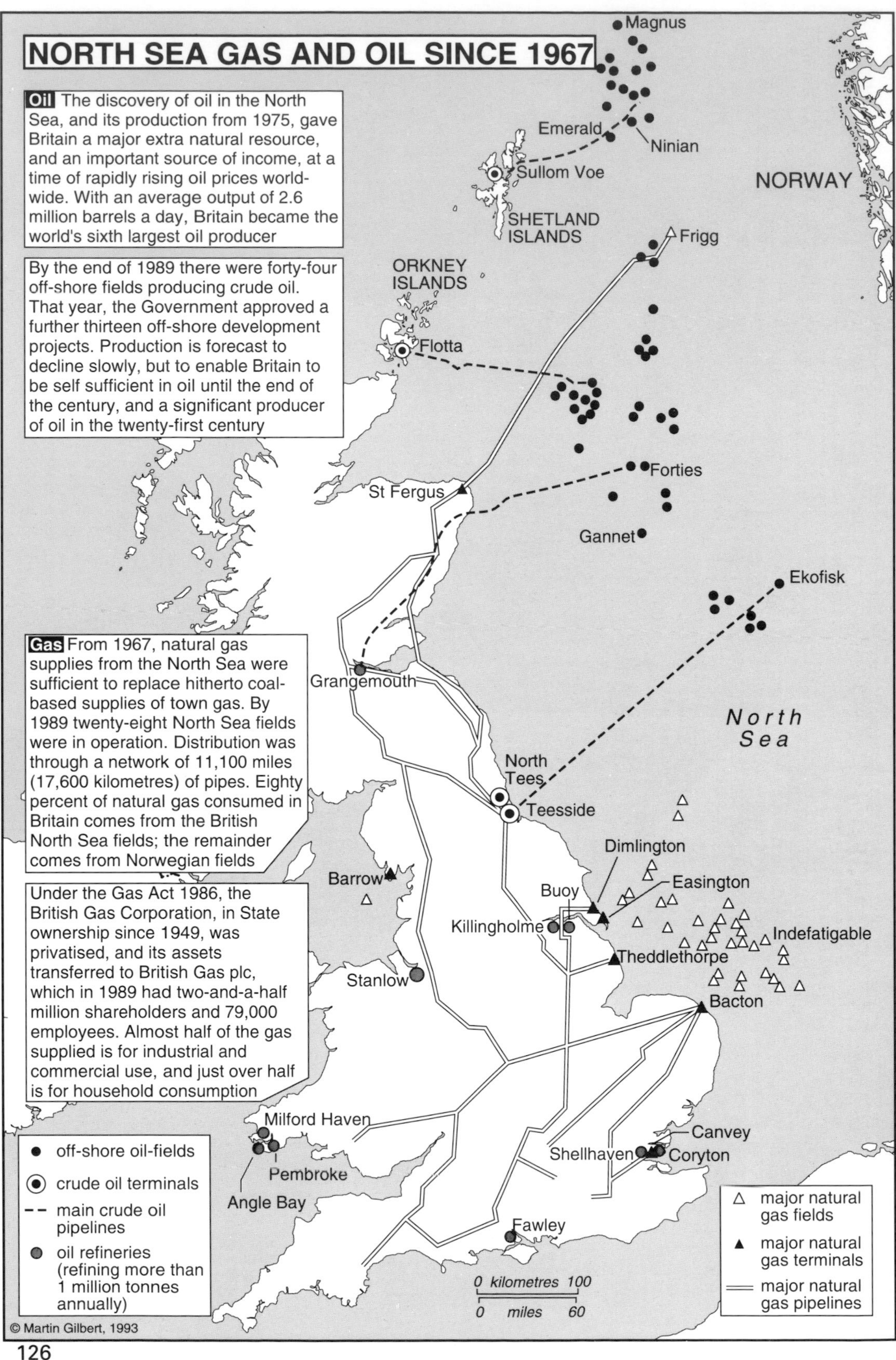
NORTH SEA GAS AND OIL SINCE 1967
Oil The discovery of oil in the North Sea, and its production from 1975, gave Britain a major extra natural resource, and an important source of income, at a time of rapidly rising oil prices world-wide. With an average output of 2.6 million barrels a day, Britain became the world's sixth largest oil producer
By the end of 1989 there were forty-four off-shore fields producing crude oil. That year, the Government approved a further thirteen off-shore development projects. Production is forecast to decline slowly, but to enable Britain to be self sufficient in oil until the end of the century, and a significant producer of oil in the twenty-first century
Gas From 1967, natural gas supplies from the North Sea were sufficient to replace hitherto coal-based supplies of town gas. By 1989 twenty-eight North Sea fields were in operation. Distribution was through a network of 11,100 miles (17,600 kilometres) of pipes. Eighty percent of natural gas consumed in Britain comes from the British North Sea fields; the remainder comes from Norwegian fields
Under the Gas Act 1986, the British Gas Corporation, in State ownership since 1949, was privatised, and its assets transferred to British Gas plc, which in 1989 had two-and-a-half million shareholders and 79,000 employees. Almost half of the gas supplied is for industrial and commercial use, and just over half is for household consumption
Magnus
Emerald
Ninian
Sullom Voe
SHETLAND ISLANDS
NORWAY
Frigg
ORKNEY ISLANDS
Flotta
Forties
St Fergus
Gannet
Ekofisk
Grangemouth
North Sea
North Tees
Teesside
Dimlington
Easington
Barrow
Buoy
Killingholme
Indefatigable
Theddlethorpe
Stanlow
Bacton
Milford Haven
Canvey
Shellhaven
Coryton
Pembroke
Angle Bay
Fawley
off-shore oil-fields
crude oil terminals
main crude oil pipelines
oil refineries (refining more than 1 million tonnes annually)
major natural gas fields
major natural gas terminals
major natural gas pipelines
0 kilometres 100
0 miles 60
© Martin Gilbert, 1993

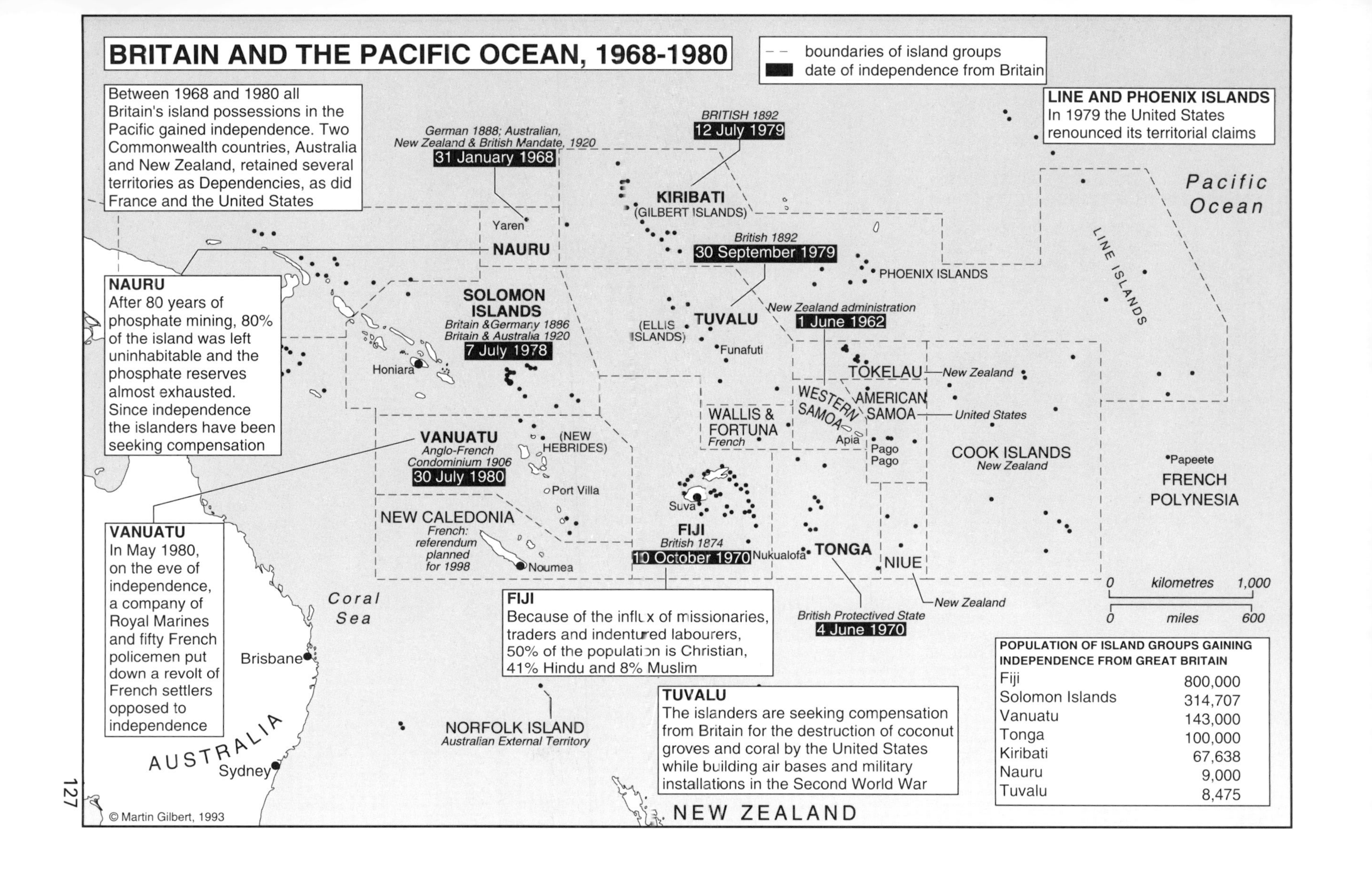

POPULATION OF ISLAND GROUPS GAINING INDEPENDENCE FROM GREAT BRITAIN	
Fiji	800,000
Solomon Islands	314,707
Vanuatu	143,000
Tonga	100,000
Kiribati	67,638
Nauru	9,000
Tuvalu	8,475

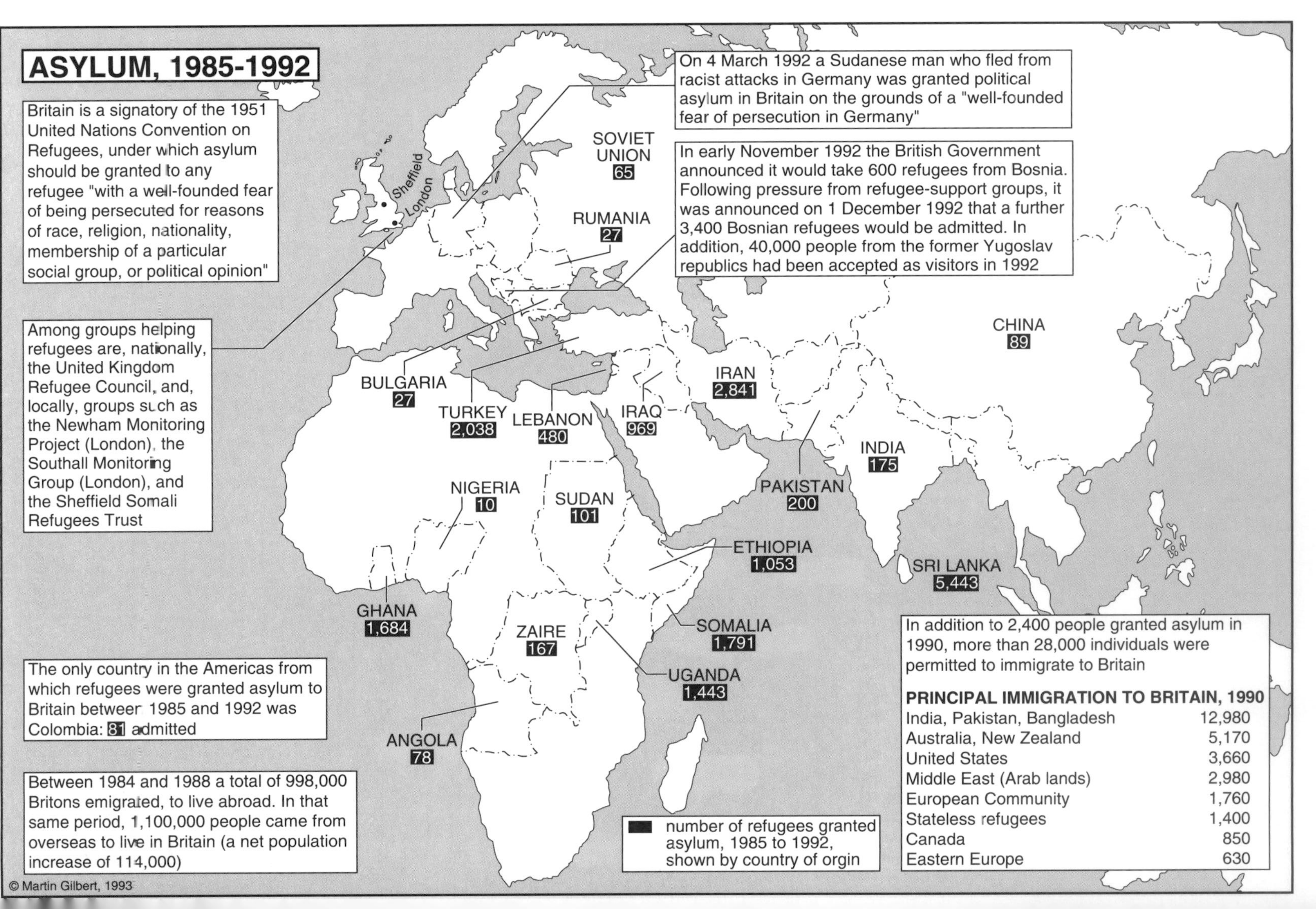
ASYLUM, 1985-1992
Britain is a signatory of the 1951 United Nations Convention on Refugees, under which asylum should be granted to any refugee "with a well-founded fear of being persecuted for reasons of race, religion, nationality, membership of a particular social group, or political opinion"
Among groups helping refugees are, nationally, the United Kingdom Refugee Council, and, locally, groups such as the Newham Monitoring Project (London), the Southall Monitoring Group (London), and the Sheffield Somali Refugees Trust
On 4 March 1992 a Sudanese man who fled from racist attacks in Germany was granted political asylum in Britain on the grounds of a "well-founded fear of persecution in Germany"
In early November 1992 the British Government announced it would take 600 refugees from Bosnia. Following pressure from refugee-support groups, it was announced on 1 December 1992 that a further 3,400 Bosnian refugees would be admitted. In addition, 40,000 people from the former Yugoslav republics had been accepted as visitors in 1992
Sheffield
London
SOVIET UNION 65
RUMANIA 27
BULGARIA 27
TURKEY 2,038
LEBANON 480
IRAQ 969
IRAN 2,841
CHINA 89
INDIA 175
PAKISTAN 200
SRI LANKA 5,443
NIGERIA 10
SUDAN 101
ETHIOPIA 1,053
GHANA 1,684
ZAIRE 167
SOMALIA 1,791
UGANDA 1,443
ANGOLA 78
The only country in the Americas from which refugees were granted asylum to Britain between 1985 and 1992 was Colombia: 81 admitted
Between 1984 and 1988 a total of 998,000 Britons emigrated, to live abroad. In that same period, 1,100,000 people came from overseas to live in Britain (a net population increase of 114,000)
number of refugees granted asylum, 1985 to 1992, shown by country of orgin
In addition to 2,400 people granted asylum in 1990, more than 28,000 individuals were permitted to immigrate to Britain
PRINCIPAL IMMIGRATION TO BRITAIN, 1990
India, Pakistan, Bangladesh 12,980
Australia, New Zealand 5,170
United States 3,660
Middle East (Arab lands) 2,980
European Community 1,760
Stateless refugees 1,400
Canada 850
Eastern Europe 630
© Martin Gilbert, 1993

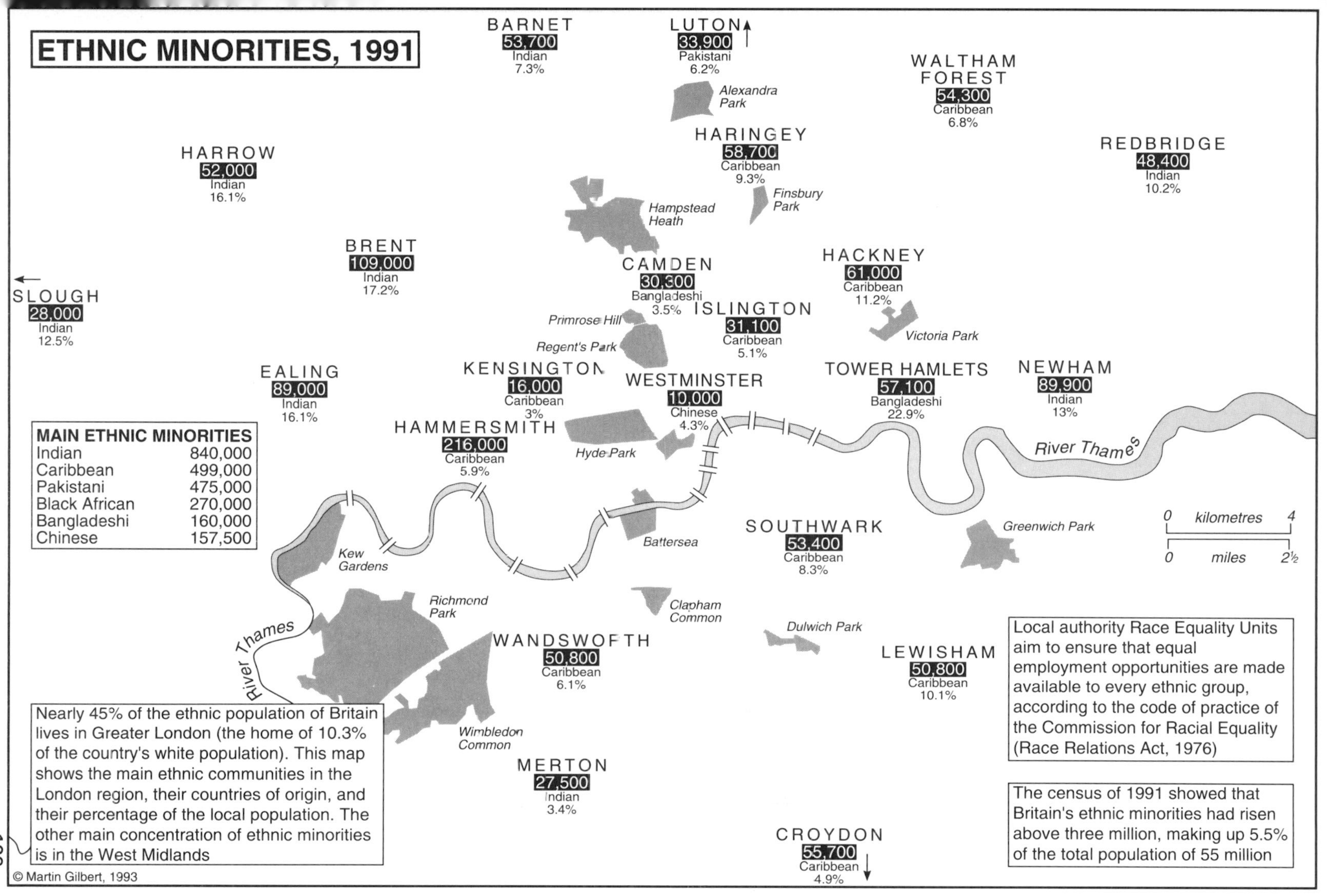
ETHNIC MINORITIES, 1991
BARNET
53,700
Indian
7.3%
LUTON
33,900
Pakistani
6.2%
Alexandra Park
WALTHAM FOREST
54,300
Caribbean
6.8%
HARINGEY
58,700
Caribbean
9.3%
REDBRIDGE
48,400
Indian
10.2%
HARROW
52,000
Indian
16.1%
Finsbury Park
Hampstead Heath
BRENT
109,000
Indian
17.2%
CAMDEN
30,300
Bangladeshi
3.5%
HACKNEY
61,000
Caribbean
11.2%
SLOUGH
28,000
Indian
12.5%
ISLINGTON
31,100
Caribbean
5.1%
Primrose Hill
Victoria Park
Regent's Park
EALING
89,000
Indian
16.1%
KENSINGTON
16,000
Caribbean
3%
WESTMINSTER
10,000
Chinese
4.3%
TOWER HAMLETS
57,100
Bangladeshi
22.9%
NEWHAM
89,900
Indian
13%
HAMMERSMITH
216,000
Caribbean
5.9%
Hyde Park
River Thames
MAIN ETHNIC MINORITIES
Indian 840,000
Caribbean 499,000
Pakistani 475,000
Black African 270,000
Bangladeshi 160,000
Chinese 157,500
SOUTHWARK
53,400
Caribbean
8.3%
Greenwich Park
0 kilometres 4
0 miles 2½
Kew Gardens
Battersea
Richmond Park
Clapham Common
River Thames
Dulwich Park
WANDSWORTH
50,800
Caribbean
6.1%
LEWISHAM
50,800
Caribbean
10.1%
Local authority Race Equality Units aim to ensure that equal employment opportunities are made available to every ethnic group, according to the code of practice of the Commission for Racial Equality (Race Relations Act, 1976)
Nearly 45% of the ethnic population of Britain lives in Greater London (the home of 10.3% of the country's white population). This map shows the main ethnic communities in the London region, their countries of origin, and their percentage of the local population. The other main concentration of ethnic minorities is in the West Midlands
Wimbledon Common
MERTON
27,500
Indian
3.4%
The census of 1991 showed that Britain's ethnic minorities had risen above three million, making up 5.5% of the total population of 55 million
CROYDON
55,700
Caribbean
4.9%
© Martin Gilbert, 1993

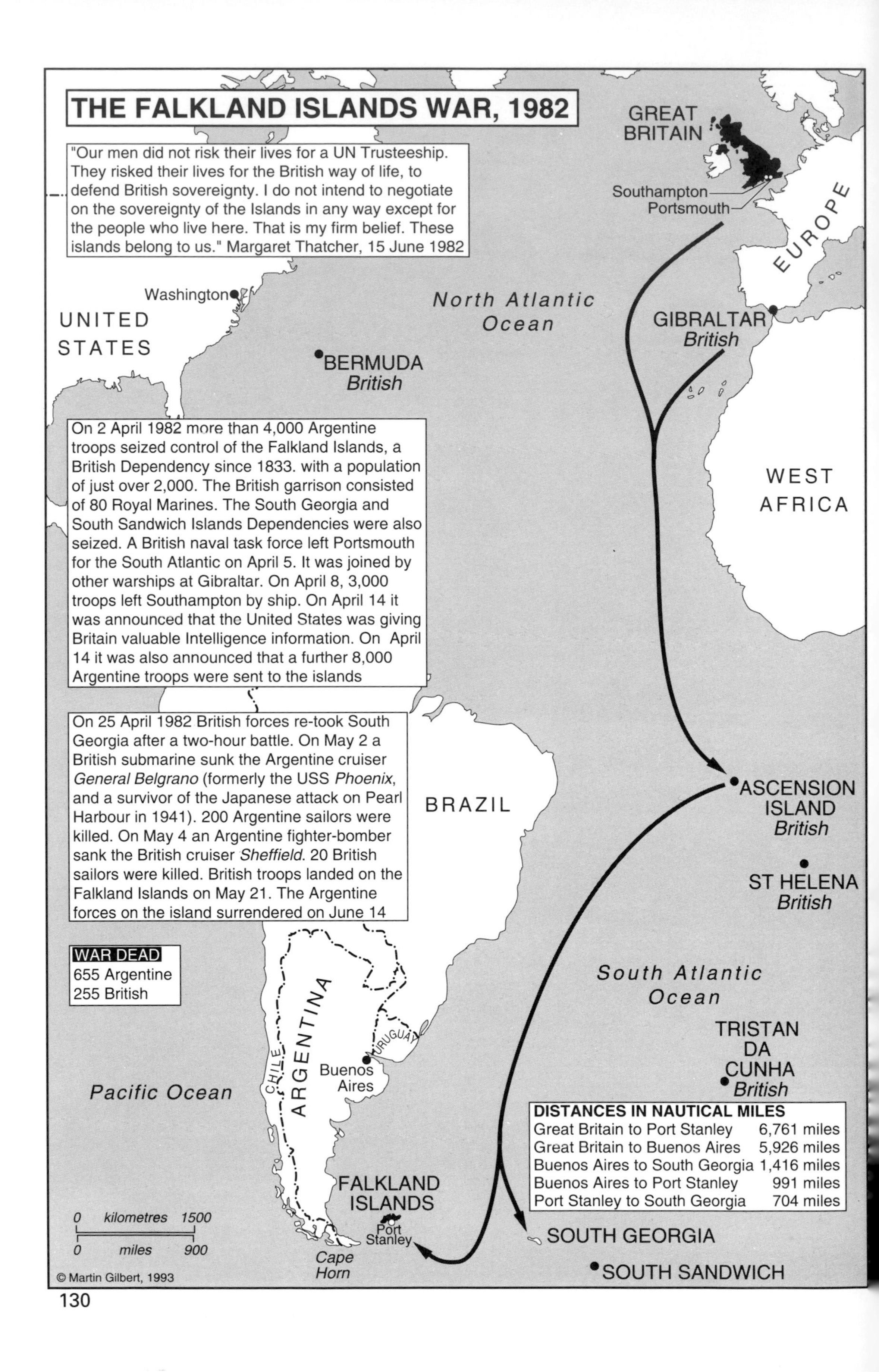
THE FALKLAND ISLANDS WAR, 1982
"Our men did not risk their lives for a UN Trusteeship. They risked their lives for the British way of life, to defend British sovereignty. I do not intend to negotiate on the sovereignty of the Islands in any way except for the people who live here. That is my firm belief. These islands belong to us." Margaret Thatcher, 15 June 1982
GREAT BRITAIN
Southampton
Portsmouth
EUROPE
Washington
UNITED STATES
North Atlantic Ocean
GIBRALTAR
British
BERMUDA
British
On 2 April 1982 more than 4,000 Argentine troops seized control of the Falkland Islands, a British Dependency since 1833. with a population of just over 2,000. The British garrison consisted of 80 Royal Marines. The South Georgia and South Sandwich Islands Dependencies were also seized. A British naval task force left Portsmouth for the South Atlantic on April 5. It was joined by other warships at Gibraltar. On April 8, 3,000 troops left Southampton by ship. On April 14 it was announced that the United States was giving Britain valuable Intelligence information. On April 14 it was also announced that a further 8,000 Argentine troops were sent to the islands
WEST AFRICA
On 25 April 1982 British forces re-took South Georgia after a two-hour battle. On May 2 a British submarine sunk the Argentine cruiser *General Belgrano* (formerly the USS *Phoenix*, and a survivor of the Japanese attack on Pearl Harbour in 1941). 200 Argentine sailors were killed. On May 4 an Argentine fighter-bomber sank the British cruiser *Sheffield*. 20 British sailors were killed. British troops landed on the Falkland Islands on May 21. The Argentine forces on the island surrendered on June 14
BRAZIL
ASCENSION ISLAND
British
ST HELENA
British
WAR DEAD
655 Argentine
255 British
ARGENTINA
CHILE
URUGUAY
Buenos Aires
South Atlantic Ocean
TRISTAN DA CUNHA
British
Pacific Ocean
DISTANCES IN NAUTICAL MILES
Great Britain to Port Stanley 6,761 miles
Great Britain to Buenos Aires 5,926 miles
Buenos Aires to South Georgia 1,416 miles
Buenos Aires to Port Stanley 991 miles
Port Stanley to South Georgia 704 miles
FALKLAND ISLANDS
Port Stanley
SOUTH GEORGIA
0 kilometres 1500
0 miles 900
Cape Horn
SOUTH SANDWICH
© Martin Gilbert, 1993

BRITAIN, THE GULF WAR AND ITS AFTERMATH, 1990-1993

On 2 August 1990 Iraqi forces occupied Kuwait. The United Nations Security Council demanded immediate withdrawal. On 29 November 1990 the Security Council authorised UN members to use force to expel Iraq from Kuwait. On 17 January 1991 Allied air forces, British among them, attacked strategic targets throughout Iraq and Iraqi-occupied Kuwait. On 24 February 1991 British forces participated in the land offensive. Four days later Iraq announced a cease-fire

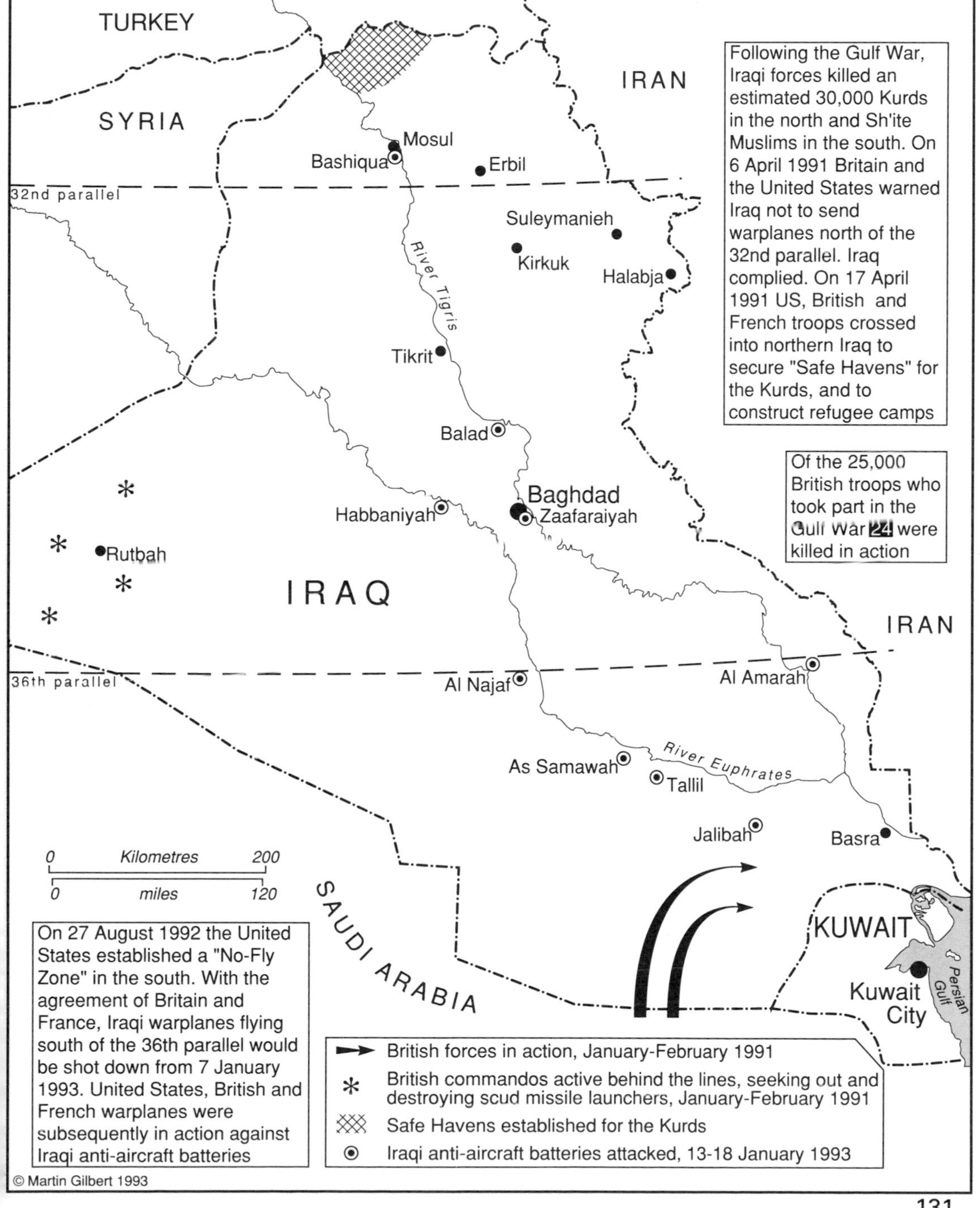

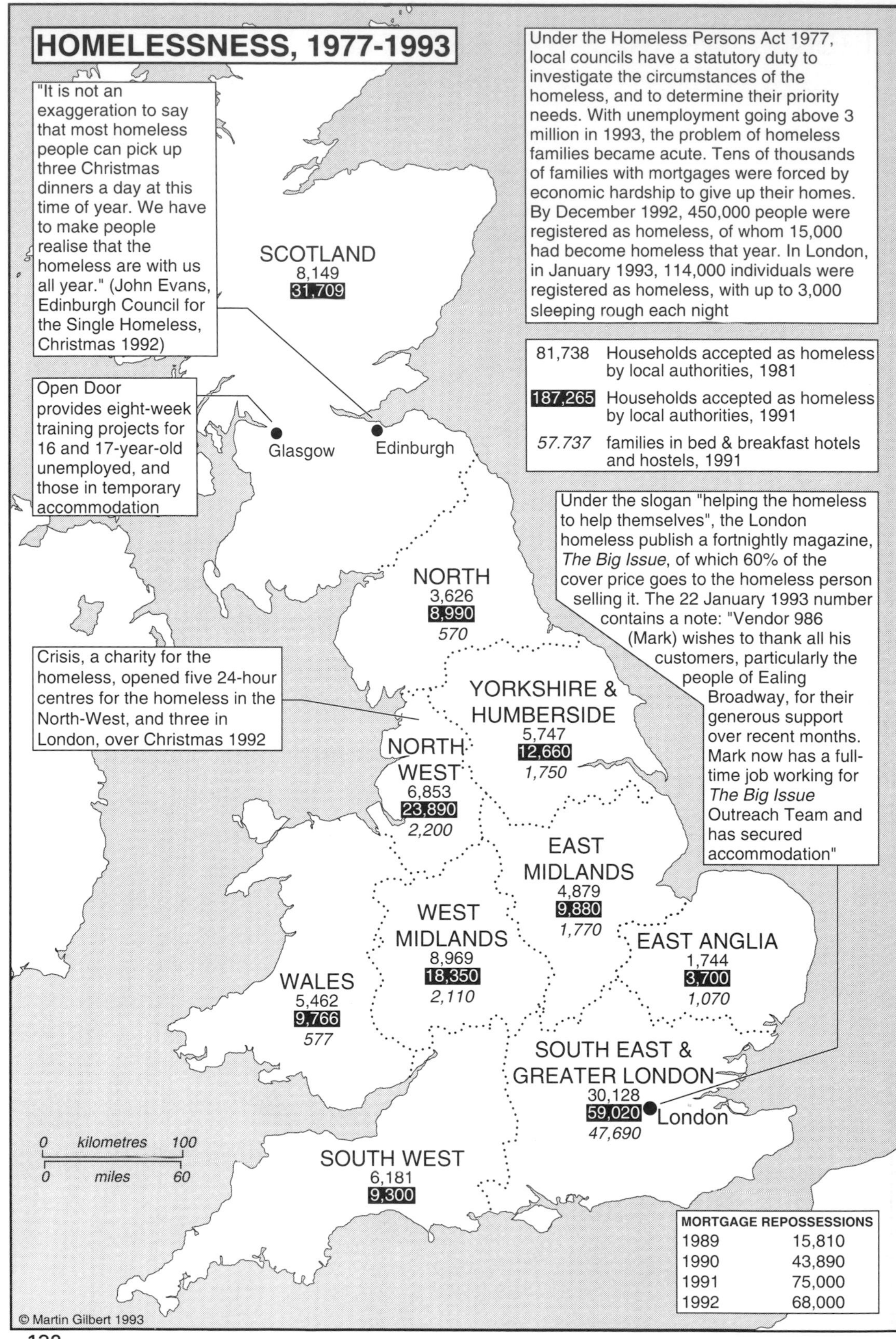
HOMELESSNESS, 1977-1993
Under the Homeless Persons Act 1977, local councils have a statutory duty to investigate the circumstances of the homeless, and to determine their priority needs. With unemployment going above 3 million in 1993, the problem of homeless families became acute. Tens of thousands of families with mortgages were forced by economic hardship to give up their homes. By December 1992, 450,000 people were registered as homeless, of whom 15,000 had become homeless that year. In London, in January 1993, 114,000 individuals were registered as homeless, with up to 3,000 sleeping rough each night
"It is not an exaggeration to say that most homeless people can pick up three Christmas dinners a day at this time of year. We have to make people realise that the homeless are with us all year." (John Evans, Edinburgh Council for the Single Homeless, Christmas 1992)
81,738 Households accepted as homeless by local authorities, 1981
187,265 Households accepted as homeless by local authorities, 1991
57.737 families in bed & breakfast hotels and hostels, 1991
SCOTLAND
8,149
31,709
Open Door provides eight-week training projects for 16 and 17-year-old unemployed, and those in temporary accommodation
Glasgow
Edinburgh
Under the slogan "helping the homeless to help themselves", the London homeless publish a fortnightly magazine, The Big Issue, of which 60% of the cover price goes to the homeless person selling it. The 22 January 1993 number contains a note: "Vendor 986 (Mark) wishes to thank all his customers, particularly the people of Ealing Broadway, for their generous support over recent months. Mark now has a full-time job working for The Big Issue Outreach Team and has secured accommodation"
NORTH
3,626
8,990
570
Crisis, a charity for the homeless, opened five 24-hour centres for the homeless in the North-West, and three in London, over Christmas 1992
YORKSHIRE & HUMBERSIDE
5,747
12,660
1,750
NORTH WEST
6,853
23,890
2,200
EAST MIDLANDS
4,879
9,880
1,770
WEST MIDLANDS
8,969
18,350
2,110
EAST ANGLIA
1,744
3,700
1,070
WALES
5,462
9,766
577
SOUTH EAST & GREATER LONDON
30,128
59,020
47,690
London
0 kilometres 100
0 miles 60
SOUTH WEST
6,181
9,300
MORTGAGE REPOSSESSIONS
1989 15,810
1990 43,890
1991 75,000
1992 68,000
© Martin Gilbert 1993

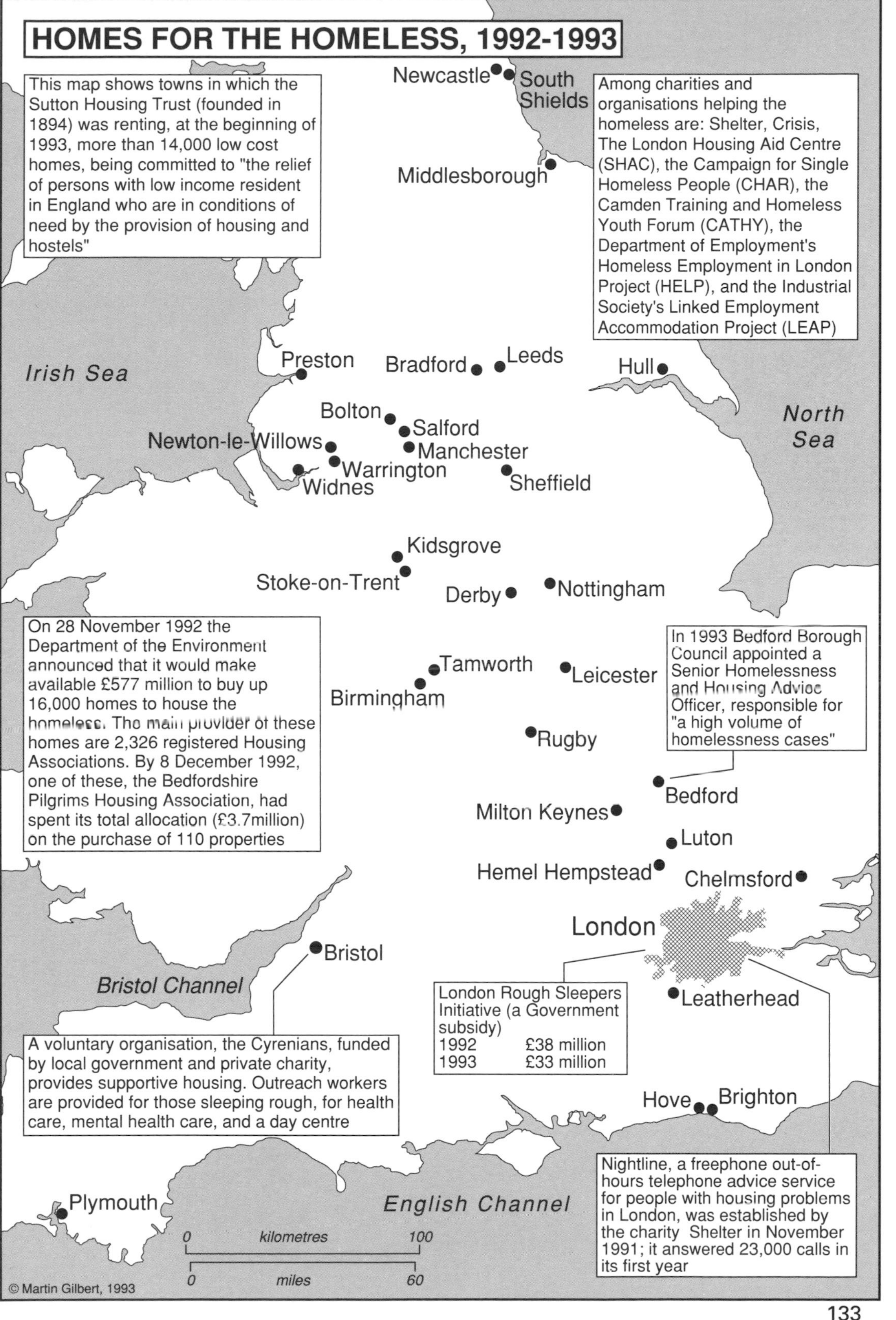
HOMES FOR THE HOMELESS, 1992-1993
This map shows towns in which the Sutton Housing Trust (founded in 1894) was renting, at the beginning of 1993, more than 14,000 low cost homes, being committed to "the relief of persons with low income resident in England who are in conditions of need by the provision of housing and hostels"
Among charities and organisations helping the homeless are: Shelter, Crisis, The London Housing Aid Centre (SHAC), the Campaign for Single Homeless People (CHAR), the Camden Training and Homeless Youth Forum (CATHY), the Department of Employment's Homeless Employment in London Project (HELP), and the Industrial Society's Linked Employment Accommodation Project (LEAP)
Newcastle
South Shields
Middlesborough
Irish Sea
Preston
Bradford
Leeds
Hull
North Sea
Bolton
Salford
Manchester
Newton-le-Willows
Warrington
Widnes
Sheffield
Kidsgrove
Stoke-on-Trent
Derby
Nottingham
On 28 November 1992 the Department of the Environment announced that it would make available £577 million to buy up 16,000 homes to house the homeless. The main provider of these homes are 2,326 registered Housing Associations. By 8 December 1992, one of these, the Bedfordshire Pilgrims Housing Association, had spent its total allocation (£3.7million) on the purchase of 110 properties
In 1993 Bedford Borough Council appointed a Senior Homelessness and Housing Advice Officer, responsible for "a high volume of homelessness cases"
Tamworth
Birmingham
Leicester
Rugby
Bedford
Milton Keynes
Luton
Hemel Hempstead
Chelmsford
London
Bristol
Bristol Channel
London Rough Sleepers Initiative (a Government subsidy)
1992 £38 million
1993 £33 million
Leatherhead
A voluntary organisation, the Cyrenians, funded by local government and private charity, provides supportive housing. Outreach workers are provided for those sleeping rough, for health care, mental health care, and a day centre
Hove
Brighton
Nightline, a freephone out-of-hours telephone advice service for people with housing problems in London, was established by the charity Shelter in November 1991; it answered 23,000 calls in its first year
Plymouth
English Channel
0 kilometres 100
0 miles 60
© Martin Gilbert, 1993

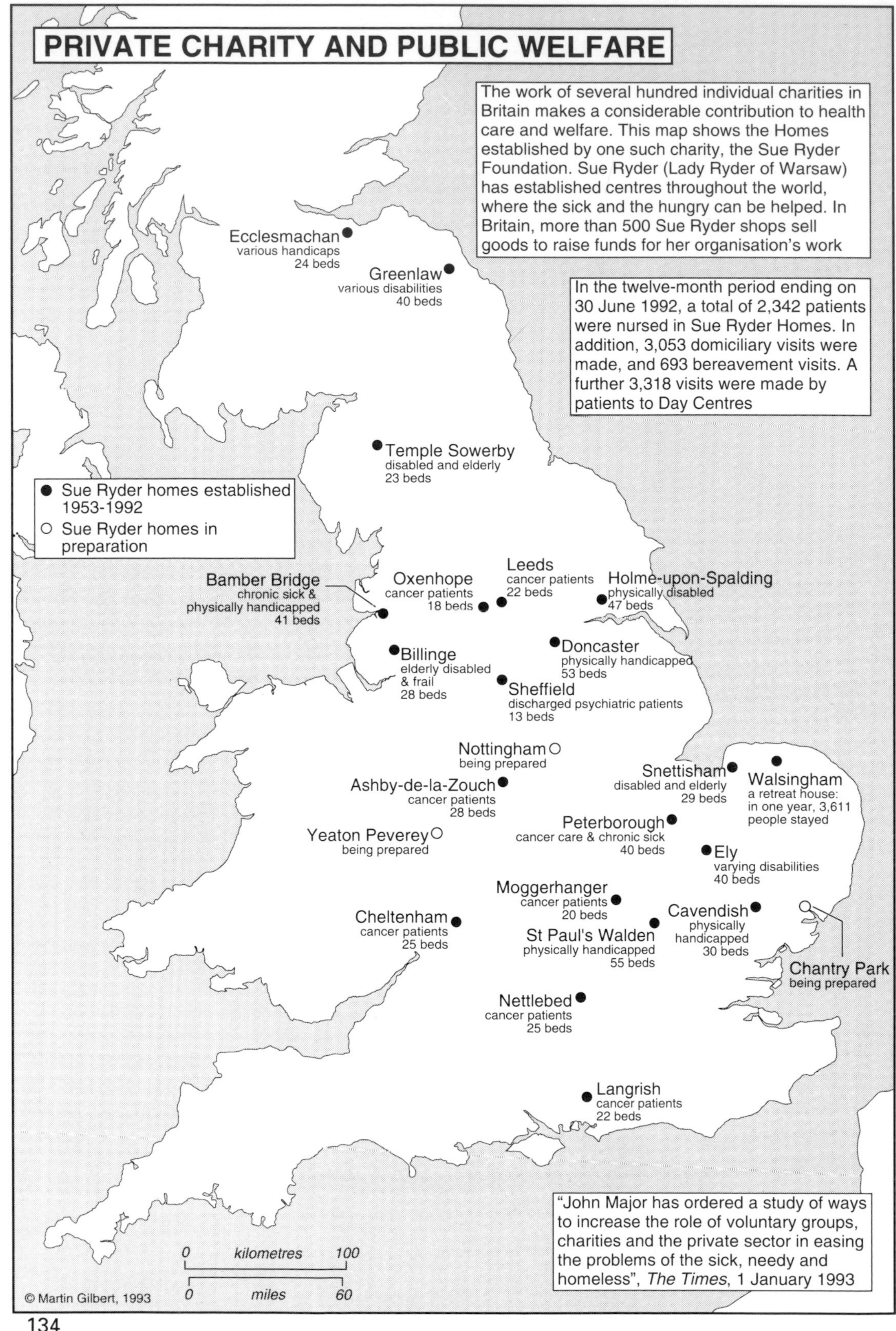
PRIVATE CHARITY AND PUBLIC WELFARE
The work of several hundred individual charities in Britain makes a considerable contribution to health care and welfare. This map shows the Homes established by one such charity, the Sue Ryder Foundation. Sue Ryder (Lady Ryder of Warsaw) has established centres throughout the world, where the sick and the hungry can be helped. In Britain, more than 500 Sue Ryder shops sell goods to raise funds for her organisation's work
In the twelve-month period ending on 30 June 1992, a total of 2,342 patients were nursed in Sue Ryder Homes. In addition, 3,053 domiciliary visits were made, and 693 bereavement visits. A further 3,318 visits were made by patients to Day Centres
Sue Ryder homes established 1953-1992
Sue Ryder homes in preparation
Ecclesmachan various handicaps 24 beds
Greenlaw various disabilities 40 beds
Temple Sowerby disabled and elderly 23 beds
Bamber Bridge chronic sick & physically handicapped 41 beds
Oxenhope cancer patients 18 beds
Leeds cancer patients 22 beds
Holme-upon-Spalding physically disabled 47 beds
Billinge elderly disabled & frail 28 beds
Doncaster physically handicapped 53 beds
Sheffield discharged psychiatric patients 13 beds
Nottingham being prepared
Ashby-de-la-Zouch cancer patients 28 beds
Snettisham disabled and elderly 29 beds
Walsingham a retreat house: in one year, 3,611 people stayed
Peterborough cancer care & chronic sick 40 beds
Yeaton Peverey being prepared
Ely varying disabilities 40 beds
Moggerhanger cancer patients 20 beds
Cheltenham cancer patients 25 beds
Cavendish physically handicapped 30 beds
St Paul's Walden physically handicapped 55 beds
Chantry Park being prepared
Nettlebed cancer patients 25 beds
Langrish cancer patients 22 beds
"John Major has ordered a study of ways to increase the role of voluntary groups, charities and the private sector in easing the problems of the sick, needy and homeless", The Times, 1 January 1993
0 kilometres 100
0 miles 60
© Martin Gilbert, 1993

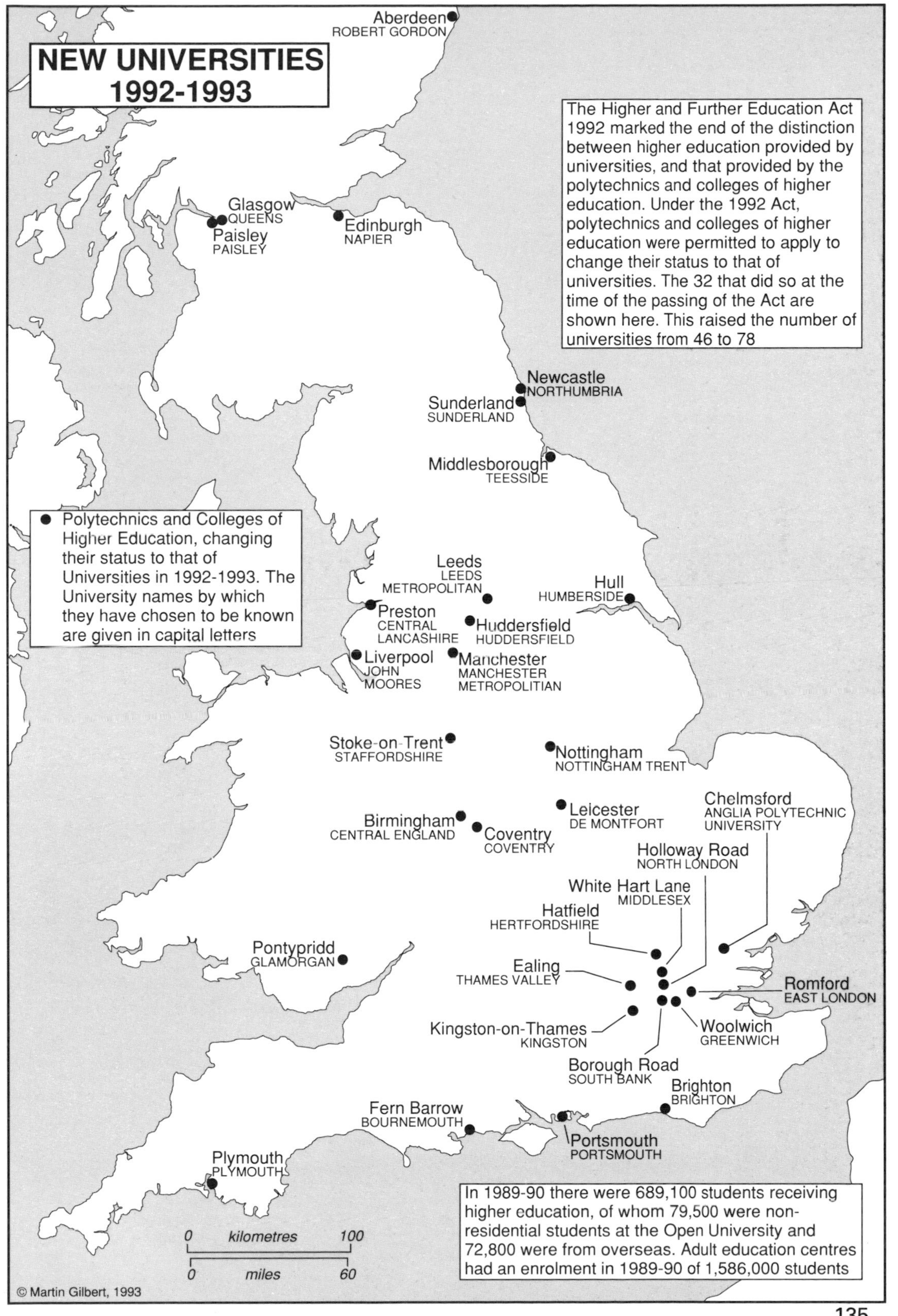
NEW UNIVERSITIES
1992-1993
The Higher and Further Education Act 1992 marked the end of the distinction between higher education provided by universities, and that provided by the polytechnics and colleges of higher education. Under the 1992 Act, polytechnics and colleges of higher education were permitted to apply to change their status to that of universities. The 32 that did so at the time of the passing of the Act are shown here. This raised the number of universities from 46 to 78
Polytechnics and Colleges of Higher Education, changing their status to that of Universities in 1992-1993. The University names by which they have chosen to be known are given in capital letters
In 1989-90 there were 689,100 students receiving higher education, of whom 79,500 were non-residential students at the Open University and 72,800 were from overseas. Adult education centres had an enrolment in 1989-90 of 1,586,000 students
Aberdeen
ROBERT GORDON
Glasgow
QUEENS
Paisley
PAISLEY
Edinburgh
NAPIER
Newcastle
NORTHUMBRIA
Sunderland
SUNDERLAND
Middlesborough
TEESSIDE
Leeds
LEEDS METROPOLITAN
Hull
HUMBERSIDE
Preston
CENTRAL LANCASHIRE
Huddersfield
HUDDERSFIELD
Liverpool
JOHN MOORES
Manchester
MANCHESTER METROPOLITAN
Stoke-on-Trent
STAFFORDSHIRE
Nottingham
NOTTINGHAM TRENT
Chelmsford
ANGLIA POLYTECHNIC UNIVERSITY
Leicester
DE MONTFORT
Birmingham
CENTRAL ENGLAND
Coventry
COVENTRY
Holloway Road
NORTH LONDON
White Hart Lane
MIDDLESEX
Hatfield
HERTFORDSHIRE
Pontypridd
GLAMORGAN
Ealing
THAMES VALLEY
Romford
EAST LONDON
Woolwich
GREENWICH
Kingston-on-Thames
KINGSTON
Borough Road
SOUTH BANK
Brighton
BRIGHTON
Fern Barrow
BOURNEMOUTH
Portsmouth
PORTSMOUTH
Plymouth
PLYMOUTH
0 kilometres 100
0 miles 60

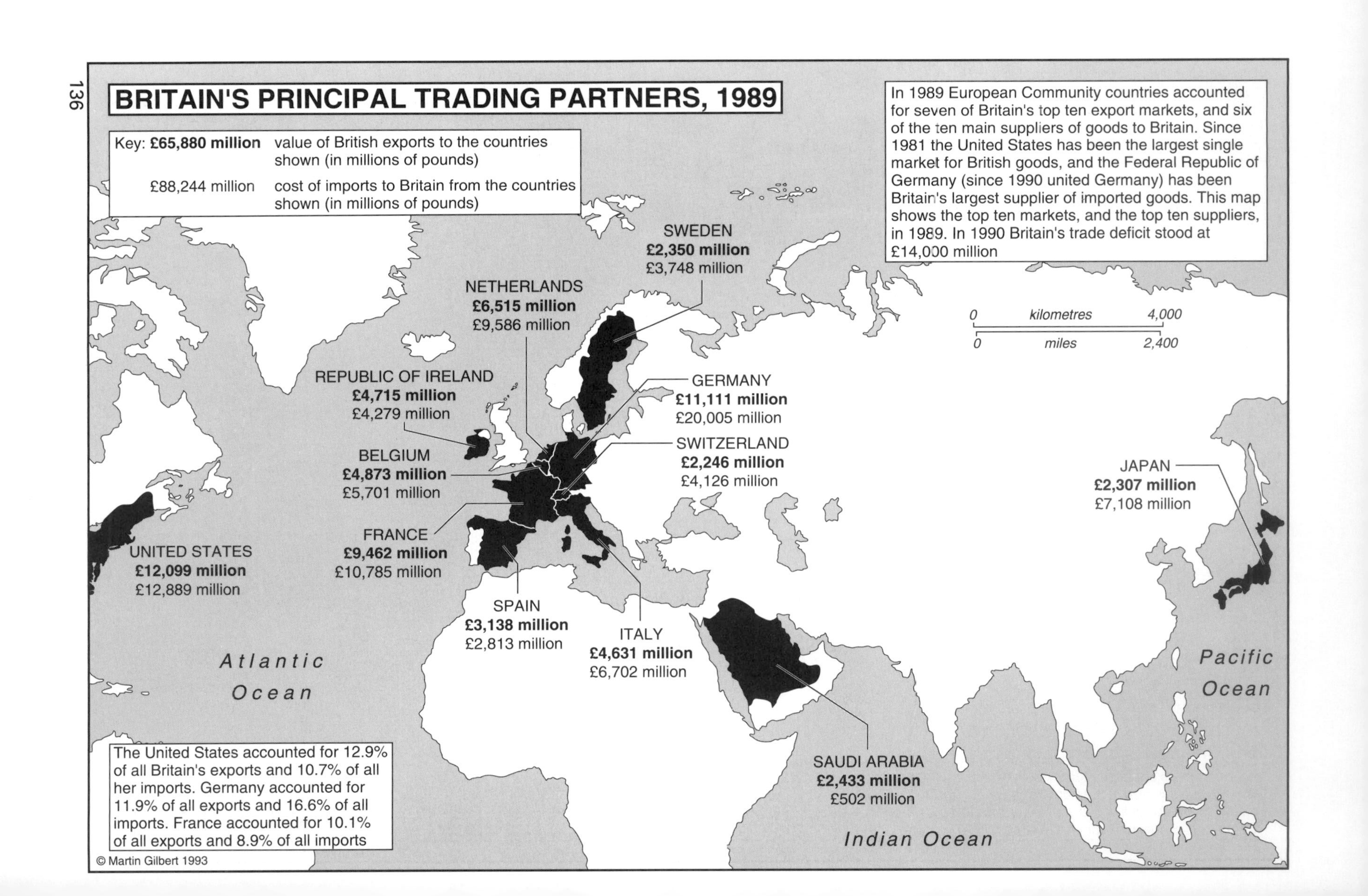
BRITAIN'S PRINCIPAL TRADING PARTNERS, 1989
Key: £65,880 million value of British exports to the countries shown (in millions of pounds)
£88,244 million cost of imports to Britain from the countries shown (in millions of pounds)
In 1989 European Community countries accounted for seven of Britain's top ten export markets, and six of the ten main suppliers of goods to Britain. Since 1981 the United States has been the largest single market for British goods, and the Federal Republic of Germany (since 1990 united Germany) has been Britain's largest supplier of imported goods. This map shows the top ten markets, and the top ten suppliers, in 1989. In 1990 Britain's trade deficit stood at £14,000 million
0 kilometres 4,000
0 miles 2,400
SWEDEN £2,350 million £3,748 million
NETHERLANDS £6,515 million £9,586 million
REPUBLIC OF IRELAND £4,715 million £4,279 million
GERMANY £11,111 million £20,005 million
SWITZERLAND £2,246 million £4,126 million
BELGIUM £4,873 million £5,701 million
JAPAN £2,307 million £7,108 million
FRANCE £9,462 million £10,785 million
UNITED STATES £12,099 million £12,889 million
SPAIN £3,138 million £2,813 million
ITALY £4,631 million £6,702 million
Atlantic Ocean
Pacific Ocean
SAUDI ARABIA £2,433 million £502 million
Indian Ocean
The United States accounted for 12.9% of all Britain's exports and 10.7% of all her imports. Germany accounted for 11.9% of all exports and 16.6% of all imports. France accounted for 10.1% of all exports and 8.9% of all imports
© Martin Gilbert 1993

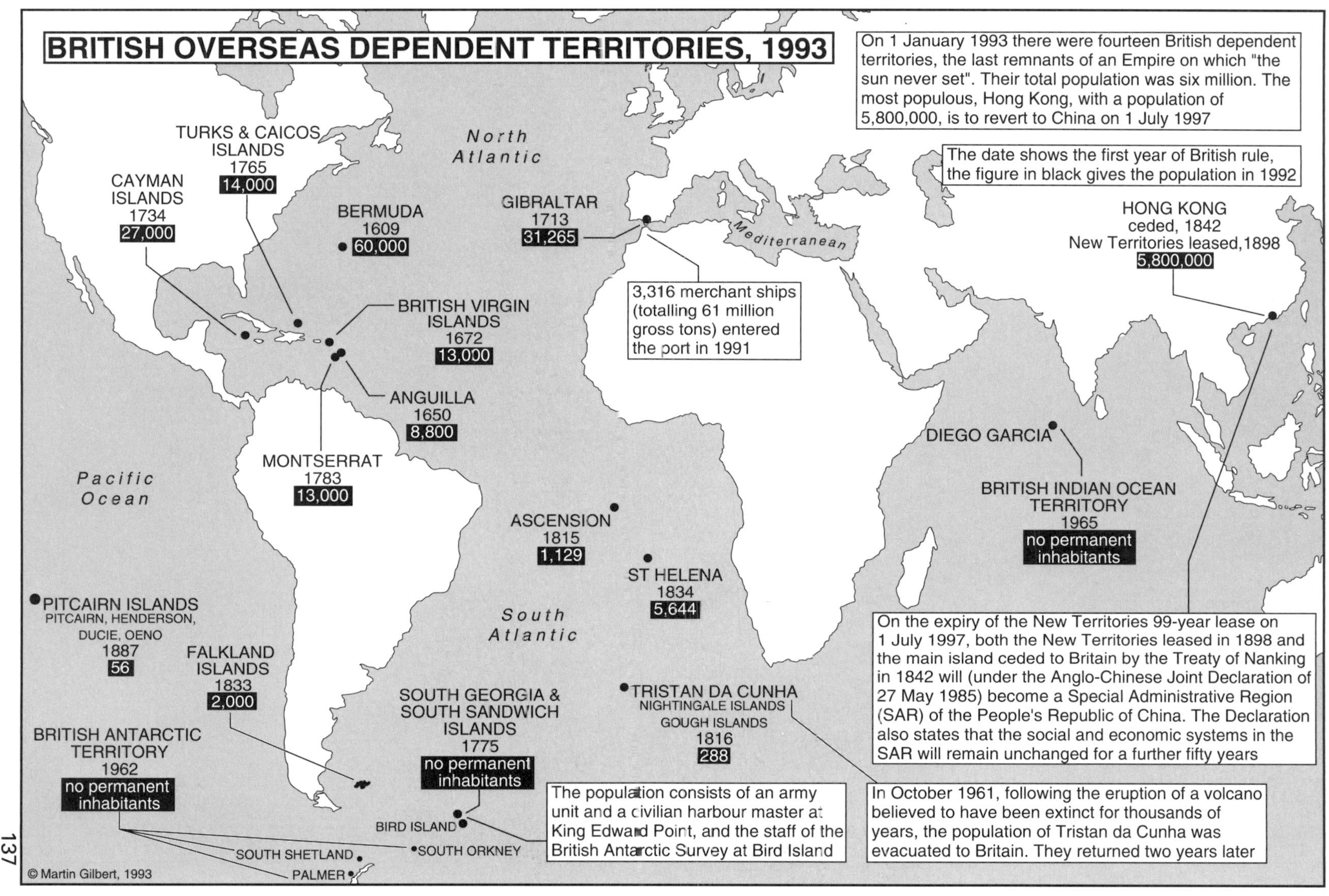
BRITISH OVERSEAS DEPENDENT TERRITORIES, 1993
On 1 January 1993 there were fourteen British dependent territories, the last remnants of an Empire on which "the sun never set". Their total population was six million. The most populous, Hong Kong, with a population of 5,800,000, is to revert to China on 1 July 1997
The date shows the first year of British rule, the figure in black gives the population in 1992
North Atlantic
TURKS & CAICOS ISLANDS 1765 14,000
CAYMAN ISLANDS 1734 27,000
BERMUDA 1609 60,000
GIBRALTAR 1713 31,265
3,316 merchant ships (totalling 61 million gross tons) entered the port in 1991
Mediterranean
HONG KONG ceded, 1842 New Territories leased,1898 5,800,000
BRITISH VIRGIN ISLANDS 1672 13,000
ANGUILLA 1650 8,800
MONTSERRAT 1783 13,000
Pacific Ocean
DIEGO GARCIA
BRITISH INDIAN OCEAN TERRITORY 1965 no permanent inhabitants
ASCENSION 1815 1,129
ST HELENA 1834 5,644
PITCAIRN ISLANDS PITCAIRN, HENDERSON, DUCIE, OENO 1887 56
South Atlantic
FALKLAND ISLANDS 1833 2,000
SOUTH GEORGIA & SOUTH SANDWICH ISLANDS 1775 no permanent inhabitants
TRISTAN DA CUNHA NIGHTINGALE ISLANDS GOUGH ISLANDS 1816 288
On the expiry of the New Territories 99-year lease on 1 July 1997, both the New Territories leased in 1898 and the main island ceded to Britain by the Treaty of Nanking in 1842 will (under the Anglo-Chinese Joint Declaration of 27 May 1985) become a Special Administrative Region (SAR) of the People's Republic of China. The Declaration also states that the social and economic systems in the SAR will remain unchanged for a further fifty years
BRITISH ANTARCTIC TERRITORY 1962 no permanent inhabitants
BIRD ISLAND
SOUTH ORKNEY
SOUTH SHETLAND
PALMER
The population consists of an army unit and a civilian harbour master at King Edward Point, and the staff of the British Antarctic Survey at Bird Island
In October 1961, following the eruption of a volcano believed to have been extinct for thousands of years, the population of Tristan da Cunha was evacuated to Britain. They returned two years later
© Martin Gilbert, 1993

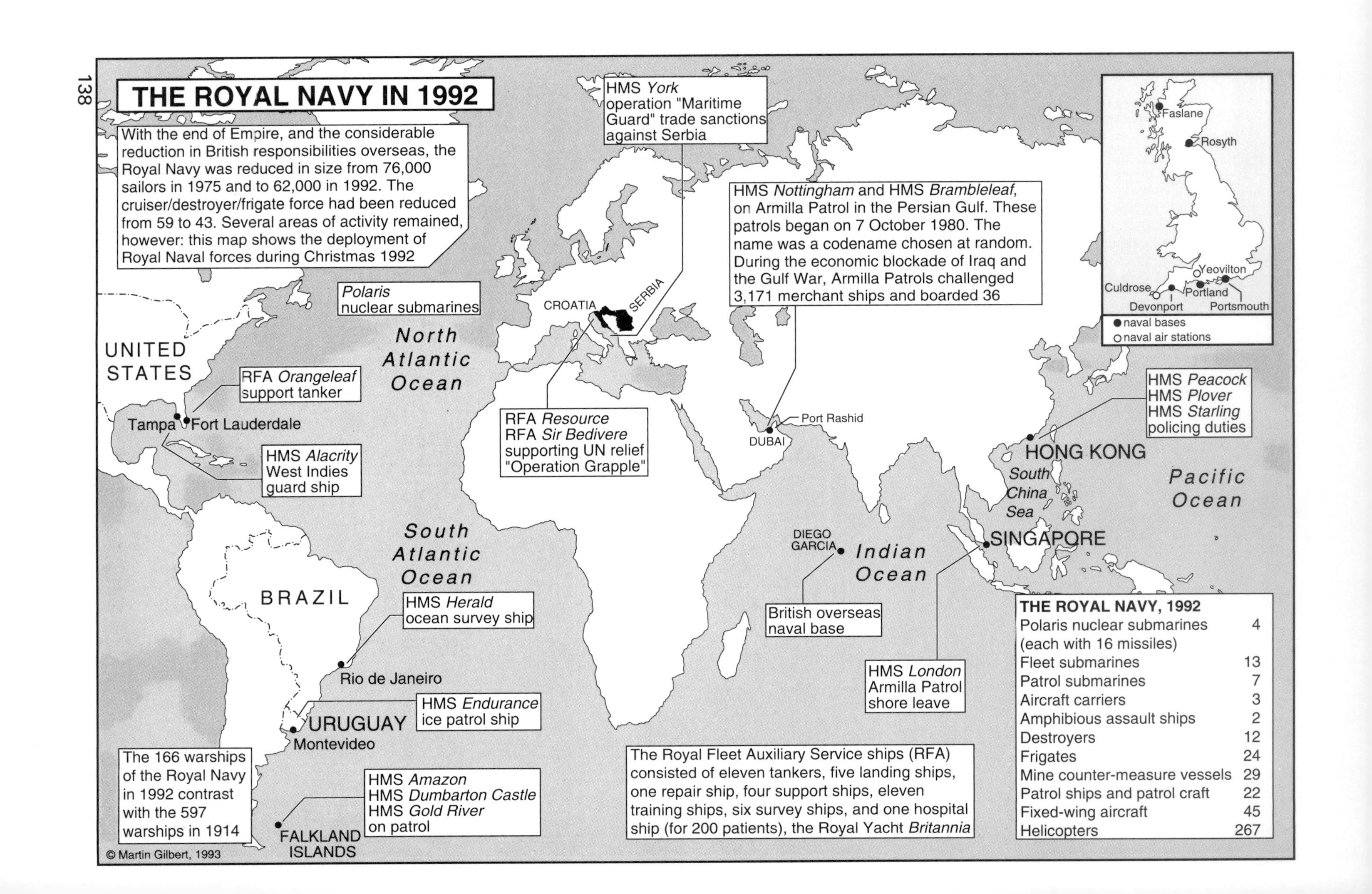
THE ROYAL NAVY IN 1992
With the end of Empire, and the considerable reduction in British responsibilities overseas, the Royal Navy was reduced in size from 76,000 sailors in 1975 and to 62,000 in 1992. The cruiser/destroyer/frigate force had been reduced from 59 to 43. Several areas of activity remained, however: this map shows the deployment of Royal Naval forces during Christmas 1992
HMS York operation "Maritime Guard" trade sanctions against Serbia
HMS Nottingham and HMS Brambleleaf, on Armilla Patrol in the Persian Gulf. These patrols began on 7 October 1980. The name was a codename chosen at random. During the economic blockade of Iraq and the Gulf War, Armilla Patrols challenged 3,171 merchant ships and boarded 36
Faslane
Rosyth
Yeovilton
Culdrose
Portland
Devonport
Portsmouth
naval bases
naval air stations
Polaris nuclear submarines
CROATIA
SERBIA
UNITED STATES
North Atlantic Ocean
RFA Orangeleaf support tanker
Tampa
Fort Lauderdale
RFA Resource
RFA Sir Bedivere
supporting UN relief "Operation Grapple"
Port Rashid
DUBAI
HMS Peacock
HMS Plover
HMS Starling
policing duties
HMS Alacrity West Indies guard ship
HONG KONG
South China Sea
Pacific Ocean
South Atlantic Ocean
DIEGO GARCIA
Indian Ocean
SINGAPORE
BRAZIL
HMS Herald ocean survey ship
British overseas naval base
Rio de Janeiro
HMS London Armilla Patrol shore leave
HMS Endurance ice patrol ship
URUGUAY
Montevideo
The 166 warships of the Royal Navy in 1992 contrast with the 597 warships in 1914
HMS Amazon
HMS Dumbarton Castle
HMS Gold River
on patrol
FALKLAND ISLANDS
The Royal Fleet Auxiliary Service ships (RFA) consisted of eleven tankers, five landing ships, one repair ship, four support ships, eleven training ships, six survey ships, and one hospital ship (for 200 patients), the Royal Yacht Britannia
THE ROYAL NAVY, 1992
Polaris nuclear submarines 4
(each with 16 missiles)
Fleet submarines 13
Patrol submarines 7
Aircraft carriers 3
Amphibious assault ships 2
Destroyers 12
Frigates 24
Mine counter-measure vessels 29
Patrol ships and patrol craft 22
Fixed-wing aircraft 45
Helicopters 267
© Martin Gilbert, 1993

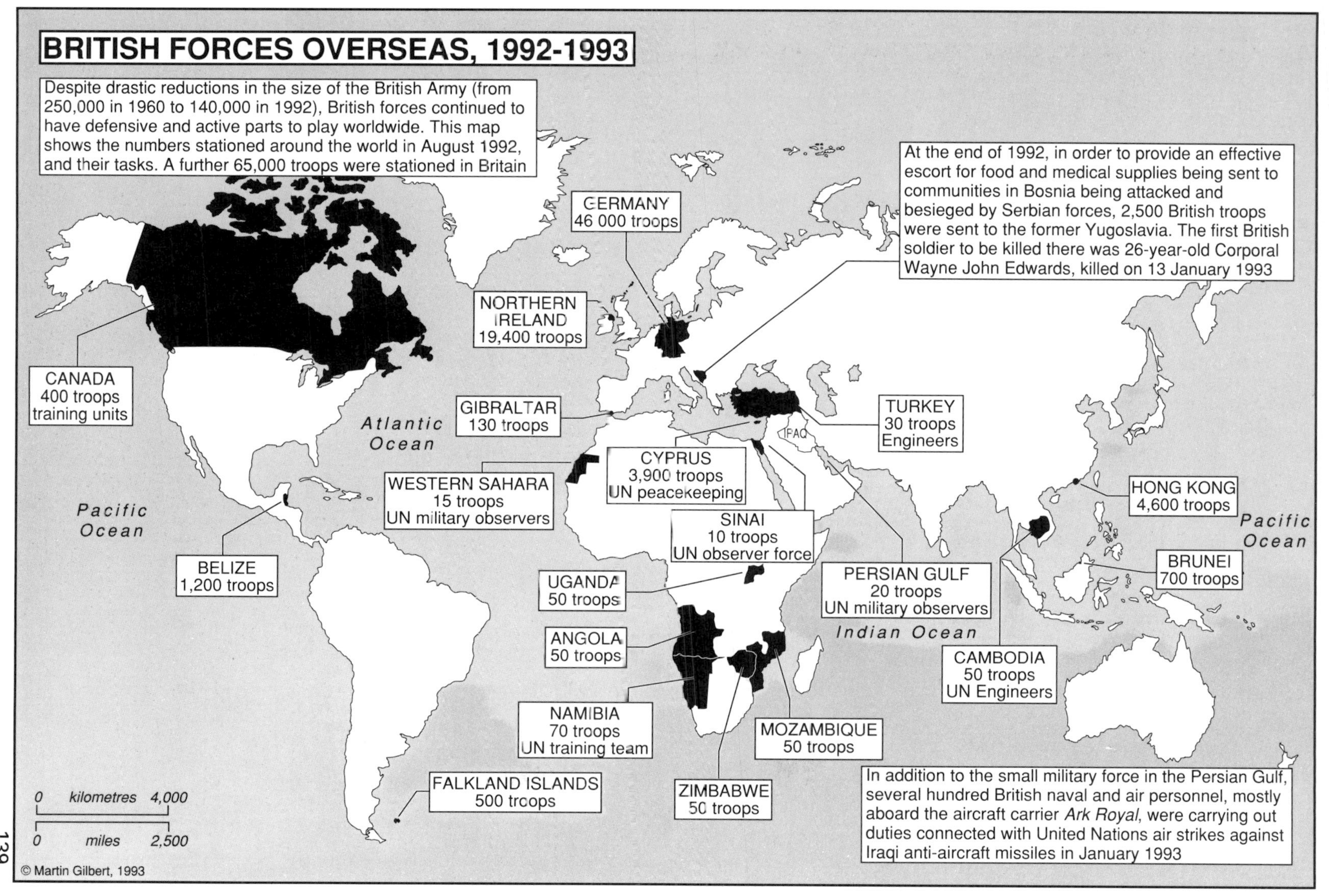
BRITISH FORCES OVERSEAS, 1992-1993
Despite drastic reductions in the size of the British Army (from 250,000 in 1960 to 140,000 in 1992), British forces continued to have defensive and active parts to play worldwide. This map shows the numbers stationed around the world in August 1992, and their tasks. A further 65,000 troops were stationed in Britain
At the end of 1992, in order to provide an effective escort for food and medical supplies being sent to communities in Bosnia being attacked and besieged by Serbian forces, 2,500 British troops were sent to the former Yugoslavia. The first British soldier to be killed there was 26-year-old Corporal Wayne John Edwards, killed on 13 January 1993
GERMANY
46 000 troops
NORTHERN IRELAND
19,400 troops
CANADA
400 troops
training units
GIBRALTAR
130 troops
Atlantic Ocean
TURKEY
30 troops
Engineers
IRAQ
CYPRUS
3,900 troops
UN peacekeeping
WESTERN SAHARA
15 troops
UN military observers
HONG KONG
4,600 troops
Pacific Ocean
SINAI
10 troops
UN observer force
Pacific Ocean
BELIZE
1,200 troops
BRUNEI
700 troops
UGANDA
50 troops
PERSIAN GULF
20 troops
UN military observers
Indian Ocean
ANGOLA
50 troops
CAMBODIA
50 troops
UN Engineers
NAMIBIA
70 troops
UN training team
MOZAMBIQUE
50 troops
FALKLAND ISLANDS
500 troops
ZIMBABWE
50 troops
In addition to the small military force in the Persian Gulf, several hundred British naval and air personnel, mostly aboard the aircraft carrier *Ark Royal*, were carrying out duties connected with United Nations air strikes against Iraqi anti-aircraft missiles in January 1993
0 kilometres 4,000
0 miles 2,500
© Martin Gilbert, 1993

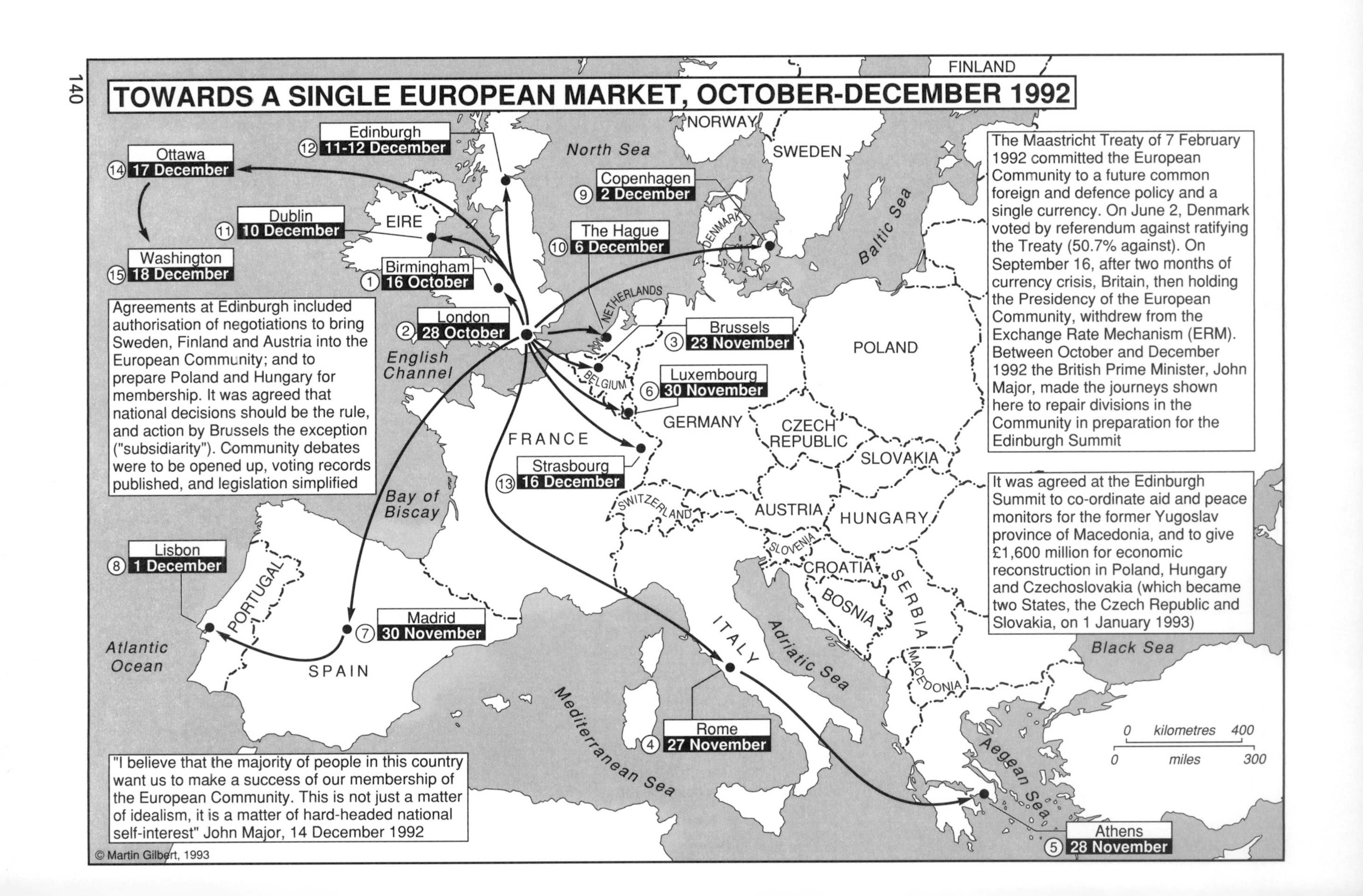

TOWARDS A SINGLE EUROPEAN MARKET, OCTOBER-DECEMBER 1992
1 Birmingham 16 October
2 London 28 October
3 Brussels 23 November
4 Rome 27 November
5 Athens 28 November
6 Luxembourg 30 November
7 Madrid 30 November
8 Lisbon 1 December
9 Copenhagen 2 December
10 The Hague 6 December
11 Dublin 10 December
12 Edinburgh 11-12 December
13 Strasbourg 16 December
14 Ottawa 17 December
15 Washington 18 December
The Maastricht Treaty of 7 February 1992 committed the European Community to a future common foreign and defence policy and a single currency. On June 2, Denmark voted by referendum against ratifying the Treaty (50.7% against). On September 16, after two months of currency crisis, Britain, then holding the Presidency of the European Community, withdrew from the Exchange Rate Mechanism (ERM). Between October and December 1992 the British Prime Minister, John Major, made the journeys shown here to repair divisions in the Community in preparation for the Edinburgh Summit
Agreements at Edinburgh included authorisation of negotiations to bring Sweden, Finland and Austria into the European Community; and to prepare Poland and Hungary for membership. It was agreed that national decisions should be the rule, and action by Brussels the exception ("subsidiarity"). Community debates were to be opened up, voting records published, and legislation simplified
It was agreed at the Edinburgh Summit to co-ordinate aid and peace monitors for the former Yugoslav province of Macedonia, and to give £1,600 million for economic reconstruction in Poland, Hungary and Czechoslovakia (which became two States, the Czech Republic and Slovakia, on 1 January 1993)
"I believe that the majority of people in this country want us to make a success of our membership of the European Community. This is not just a matter of idealism, it is a matter of hard-headed national self-interest" John Major, 14 December 1992
FINLAND
NORWAY
SWEDEN
DENMARK
NETHERLANDS
BELGIUM
GERMANY
POLAND
CZECH REPUBLIC
SLOVAKIA
AUSTRIA
HUNGARY
SWITZERLAND
SLOVENIA
CROATIA
BOSNIA
SERBIA
MACEDONIA
FRANCE
SPAIN
PORTUGAL
ITALY
EIRE
North Sea
Baltic Sea
English Channel
Bay of Biscay
Atlantic Ocean
Mediterranean Sea
Adriatic Sea
Aegean Sea
Black Sea
0 kilometres 400
0 miles 300

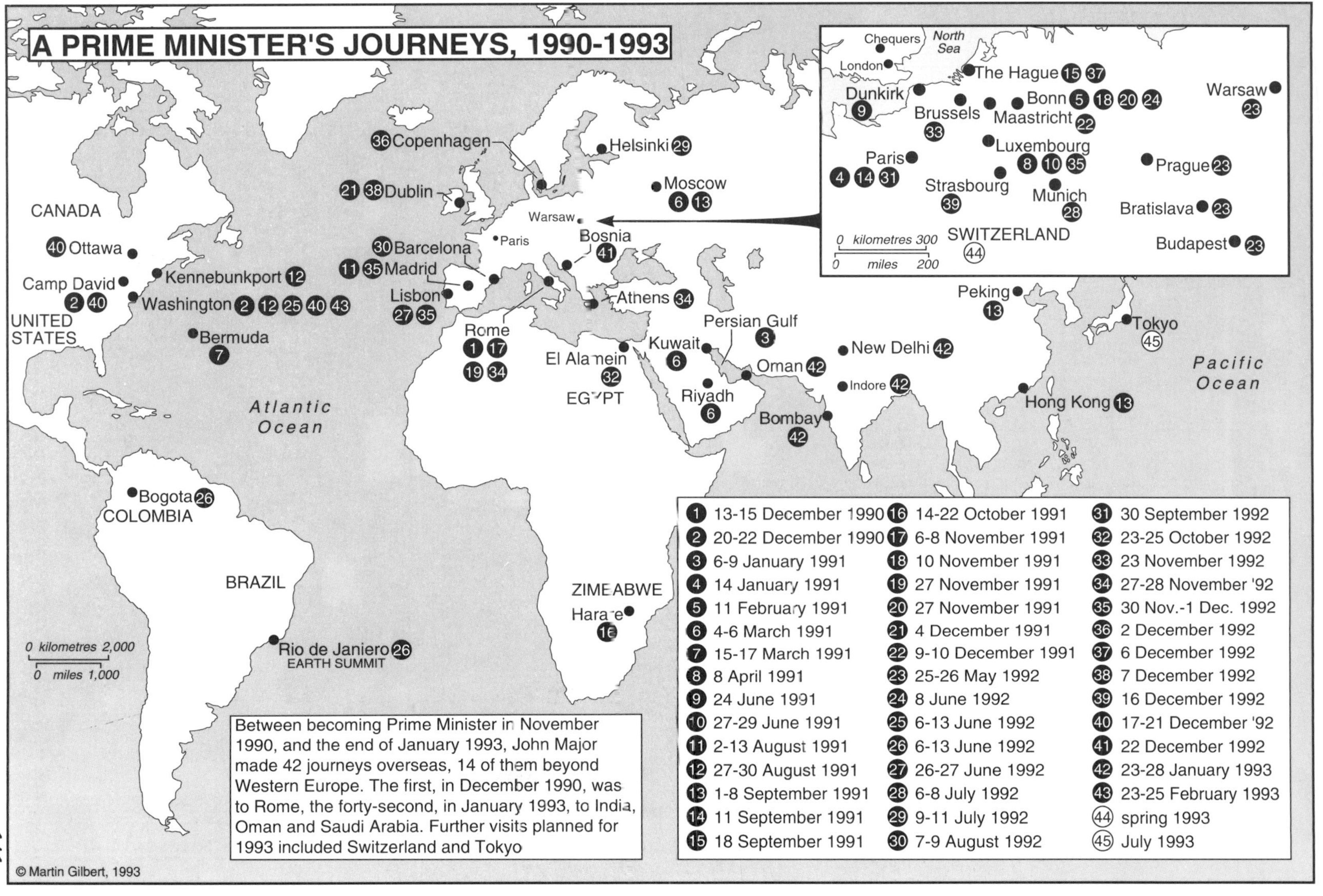
A PRIME MINISTER'S JOURNEYS, 1990-1993
CANADA
40 Ottawa
Camp David 2 40
UNITED STATES
Kennebunkport 12
Washington 2 12 25 40 43
Bermuda 7
Atlantic Ocean
36 Copenhagen
21 38 Dublin
30 Barcelona
11 35 Madrid
Lisbon 27 35
Paris
Warsaw
Rome 1 17 19 34
Bosnia 41
Athens 34
Helsinki 29
Moscow 6 13
32
Kuwait 6
Riyadh 6
Persian Gulf 3
Oman 42
New Delhi 42
Indore 42
Bombay 42
Peking 13
Hong Kong 13
Tokyo 45
Pacific Ocean
Bogota 26
COLOMBIA
BRAZIL
Rio de Janiero 26
EARTH SUMMIT
0 kilometres 2,000
0 miles 1,000
Chequers
London
North Sea
Dunkirk 9
The Hague 15 37
Brussels 33
Bonn 5 18 20 24
Maastricht 22
Luxembourg 8 10 35
Paris 4 14 31
Strasbourg 39
Munich 28
SWITZERLAND 44
Warsaw 23
Prague 23
Bratislava 23
Budapest 23
0 kilometres 300
0 miles 200
Between becoming Prime Minister in November 1990, and the end of January 1993, John Major made 42 journeys overseas, 14 of them beyond Western Europe. The first, in December 1990, was to Rome, the forty-second, in January 1993, to India, Oman and Saudi Arabia. Further visits planned for 1993 included Switzerland and Tokyo
1 13-15 December 1990
2 20-22 December 1990
3 6-9 January 1991
4 14 January 1991
5 11 February 1991
6 4-6 March 1991
7 15-17 March 1991
8 8 April 1991
9 24 June 1991
10 27-29 June 1991
11 2-13 August 1991
12 27-30 August 1991
13 1-8 September 1991
14 11 September 1991
15 18 September 1991
16 14-22 October 1991
17 6-8 November 1991
18 10 November 1991
19 27 November 1991
20 27 November 1991
21 4 December 1991
22 9-10 December 1991
23 25-26 May 1992
24 8 June 1992
25 6-13 June 1992
26 6-13 June 1992
27 26-27 June 1992
28 6-8 July 1992
29 9-11 July 1992
30 7-9 August 1992
31 30 September 1992
32 23-25 October 1992
33 23 November 1992
34 27-28 November '92
35 30 Nov.-1 Dec. 1992
36 2 December 1992
37 6 December 1992
38 7 December 1992
39 16 December 1992
40 17-21 December '92
41 22 December 1992
42 23-28 January 1993
43 23-25 February 1993
44 spring 1993
45 July 1993
© Martin Gilbert, 1993

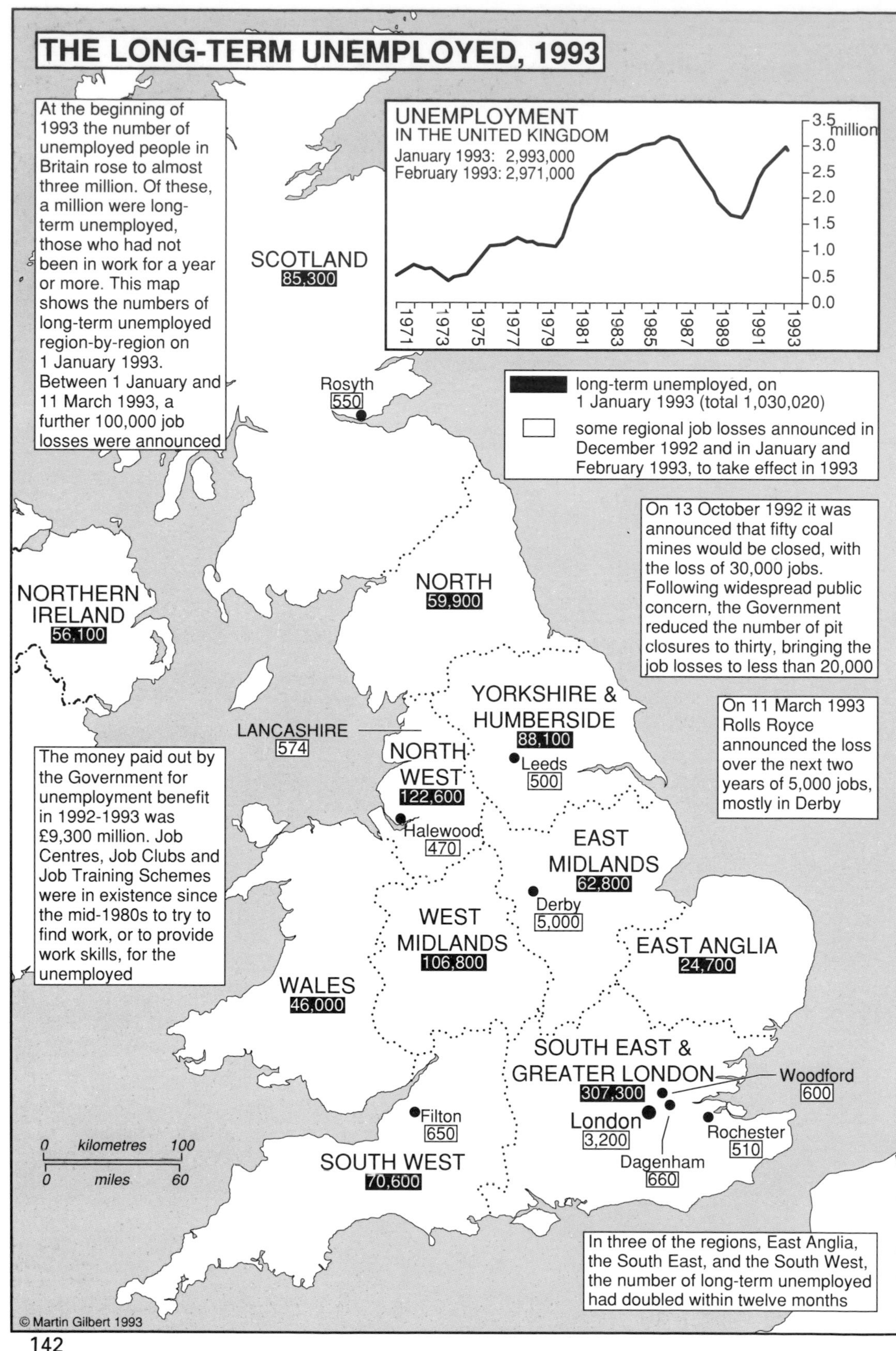
THE LONG-TERM UNEMPLOYED, 1993
At the beginning of 1993 the number of unemployed people in Britain rose to almost three million. Of these, a million were long-term unemployed, those who had not been in work for a year or more. This map shows the numbers of long-term unemployed region-by-region on 1 January 1993. Between 1 January and 11 March 1993, a further 100,000 job losses were announced
UNEMPLOYMENT
IN THE UNITED KINGDOM
January 1993: 2,993,000
February 1993: 2,971,000
3.5 million
3.0
2.5
2.0
1.5
1.0
0.5
0.0
1971
1973
1975
1977
1979
1981
1983
1985
1987
1989
1991
1993
SCOTLAND
85,300
Rosyth
550
long-term unemployed, on 1 January 1993 (total 1,030,020)
some regional job losses announced in December 1992 and in January and February 1993, to take effect in 1993
On 13 October 1992 it was announced that fifty coal mines would be closed, with the loss of 30,000 jobs. Following widespread public concern, the Government reduced the number of pit closures to thirty, bringing the job losses to less than 20,000
NORTH
59,900
NORTHERN IRELAND
56,100
YORKSHIRE & HUMBERSIDE
88,100
Leeds
500
On 11 March 1993 Rolls Royce announced the loss over the next two years of 5,000 jobs, mostly in Derby
LANCASHIRE
574
NORTH WEST
122,600
Halewood
470
The money paid out by the Government for unemployment benefit in 1992-1993 was £9,300 million. Job Centres, Job Clubs and Job Training Schemes were in existence since the mid-1980s to try to find work, or to provide work skills, for the unemployed
EAST MIDLANDS
62,800
Derby
5,000
WEST MIDLANDS
106,800
EAST ANGLIA
24,700
WALES
46,000
SOUTH EAST & GREATER LONDON
307,300
Woodford
600
London
3,200
Rochester
510
Dagenham
660
Filton
650
SOUTH WEST
70,600
0 kilometres 100
0 miles 60
In three of the regions, East Anglia, the South East, and the South West, the number of long-term unemployed had doubled within twelve months
© Martin Gilbert 1993

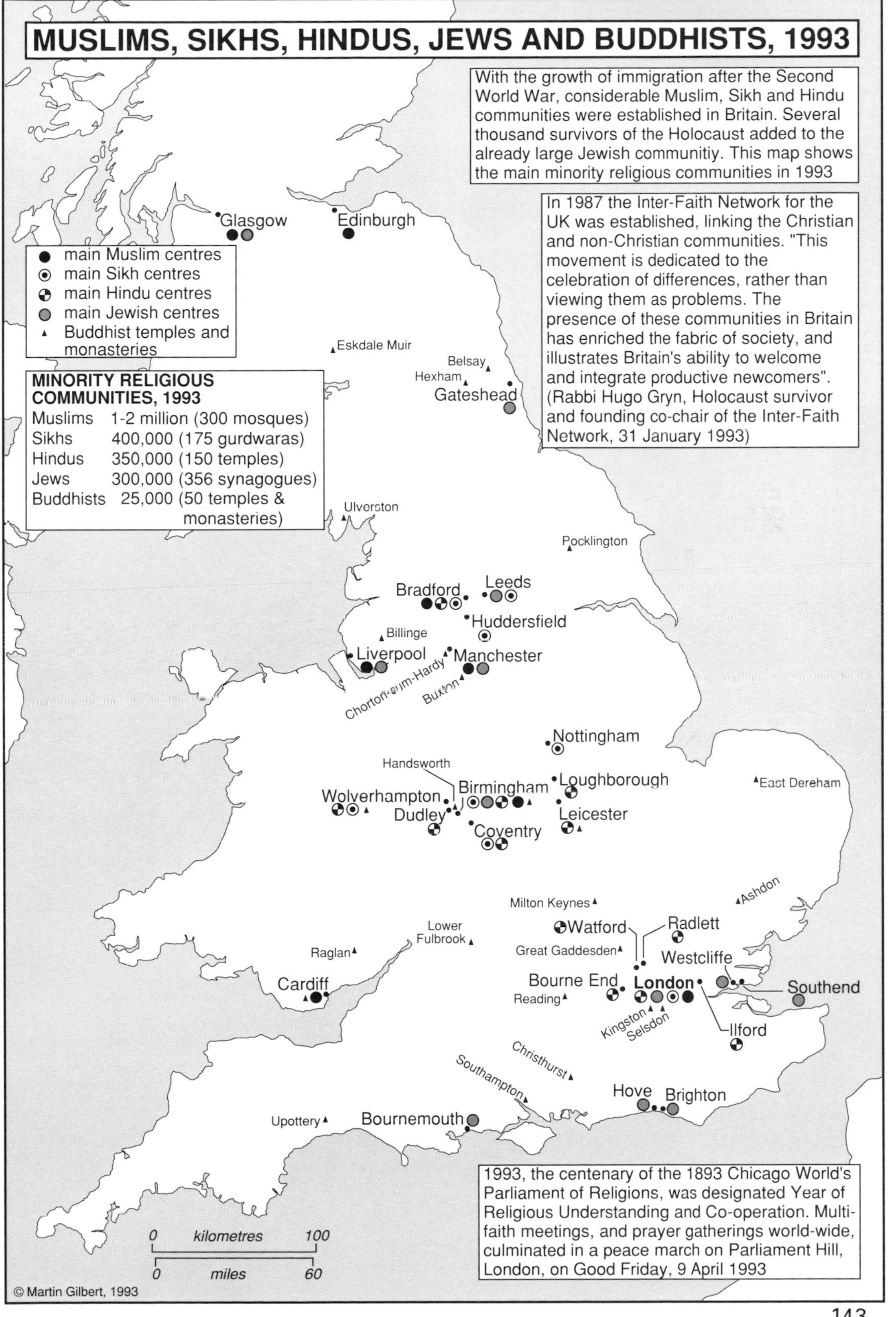

MUSLIMS, SIKHS, HINDUS, JEWS AND BUDDHISTS, 1993
With the growth of immigration after the Second World War, considerable Muslim, Sikh and Hindu communities were established in Britain. Several thousand survivors of the Holocaust added to the already large Jewish communitiy. This map shows the main minority religious communities in 1993
In 1987 the Inter-Faith Network for the UK was established, linking the Christian and non-Christian communities. "This movement is dedicated to the celebration of differences, rather than viewing them as problems. The presence of these communities in Britain has enriched the fabric of society, and illustrates Britain's ability to welcome and integrate productive newcomers". (Rabbi Hugo Gryn, Holocaust survivor and founding co-chair of the Inter-Faith Network, 31 January 1993)
main Muslim centres
main Sikh centres
main Hindu centres
main Jewish centres
Buddhist temples and monasteries
MINORITY RELIGIOUS COMMUNITIES, 1993
Muslims 1-2 million (300 mosques)
Sikhs 400,000 (175 gurdwaras)
Hindus 350,000 (150 temples)
Jews 300,000 (356 synagogues)
Buddhists 25,000 (50 temples & monasteries)
Glasgow
Edinburgh
Eskdale Muir
Belsay
Hexham
Gateshead
Ulverston
Pocklington
Bradford
Leeds
Huddersfield
Billinge
Liverpool
Manchester
Chorlton-cum-Hardy
Buxton
Nottingham
Handsworth
Birmingham
Loughborough
East Dereham
Wolverhampton
Dudley
Leicester
Coventry
Ashdon
Milton Keynes
Lower Fulbrook
Watford
Radlett
Raglan
Great Gaddesden
Westcliffe
Cardiff
Bourne End
London
Southend
Reading
Kingston
Selsdon
Ilford
Christhurst
Southampton
Hove
Brighton
Upottery
Bournemouth
1993, the centenary of the 1893 Chicago World's Parliament of Religions, was designated Year of Religious Understanding and Co-operation. Multi-faith meetings, and prayer gatherings world-wide, culminated in a peace march on Parliament Hill, London, on Good Friday, 9 April 1993
0 kilometres 100
0 miles 60
© Martin Gilbert, 1993

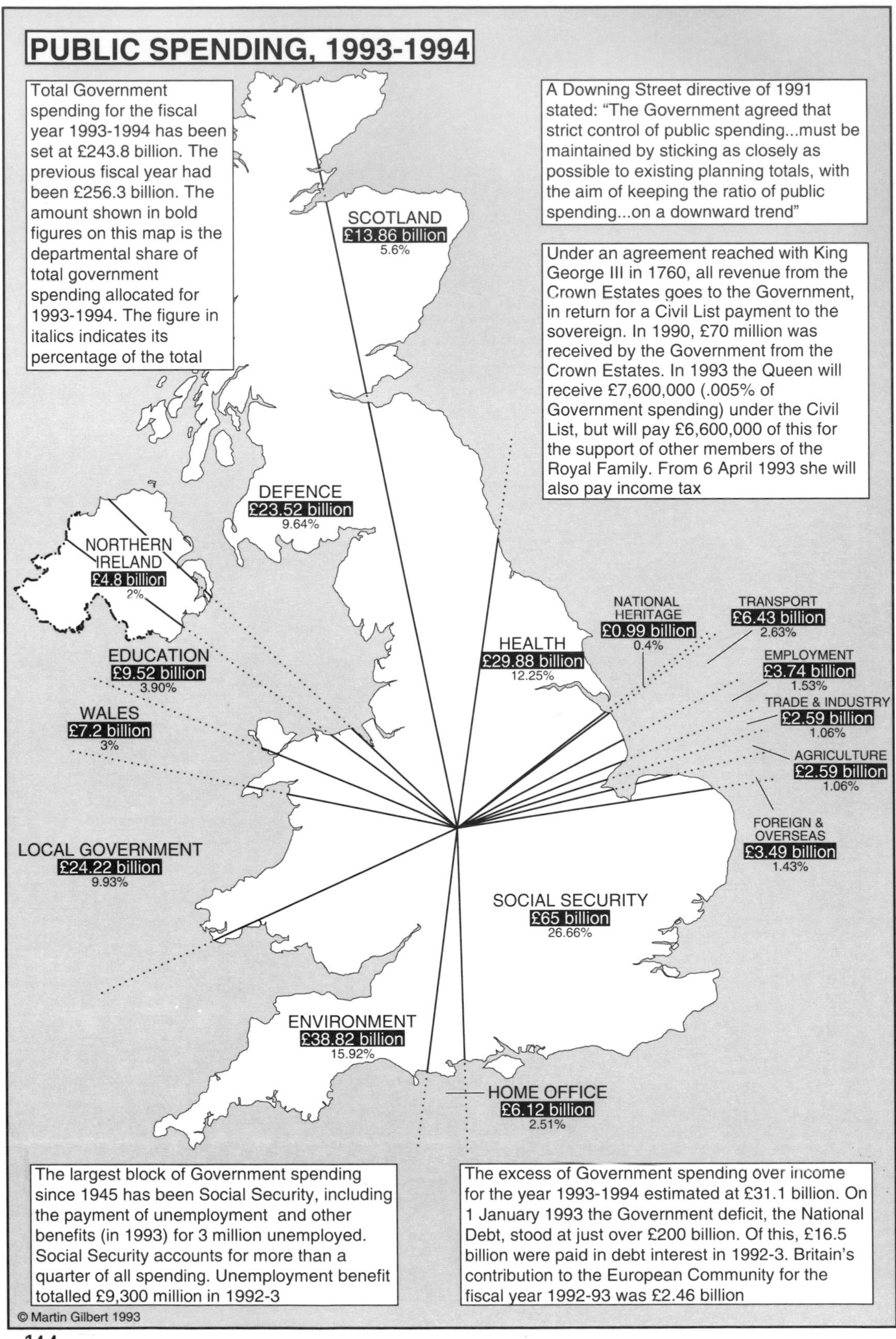
PUBLIC SPENDING, 1993-1994
Total Government spending for the fiscal year 1993-1994 has been set at £243.8 billion. The previous fiscal year had been £256.3 billion. The amount shown in bold figures on this map is the departmental share of total government spending allocated for 1993-1994. The figure in italics indicates its percentage of the total
A Downing Street directive of 1991 stated: "The Government agreed that strict control of public spending...must be maintained by sticking as closely as possible to existing planning totals, with the aim of keeping the ratio of public spending...on a downward trend"
Under an agreement reached with King George III in 1760, all revenue from the Crown Estates goes to the Government, in return for a Civil List payment to the sovereign. In 1990, £70 million was received by the Government from the Crown Estates. In 1993 the Queen will receive £7,600,000 (.005% of Government spending) under the Civil List, but will pay £6,600,000 of this for the support of other members of the Royal Family. From 6 April 1993 she will also pay income tax
SCOTLAND
£13.86 billion
5.6%
DEFENCE
£23.52 billion
9.64%
NORTHERN IRELAND
£4.8 billion
2%
EDUCATION
£9.52 billion
3.90%
WALES
£7.2 billion
3%
HEALTH
£29.88 billion
12.25%
NATIONAL HERITAGE
£0.99 billion
0.4%
TRANSPORT
£6.43 billion
2.63%
EMPLOYMENT
£3.74 billion
1.53%
TRADE & INDUSTRY
£2.59 billion
1.06%
AGRICULTURE
£2.59 billion
1.06%
FOREIGN & OVERSEAS
£3.49 billion
1.43%
LOCAL GOVERNMENT
£24.22 billion
9.93%
SOCIAL SECURITY
£65 billion
26.66%
ENVIRONMENT
£38.82 billion
15.92%
HOME OFFICE
£6.12 billion
2.51%
The largest block of Government spending since 1945 has been Social Security, including the payment of unemployment and other benefits (in 1993) for 3 million unemployed. Social Security accounts for more than a quarter of all spending. Unemployment benefit totalled £9,300 million in 1992-3
The excess of Government spending over income for the year 1993-1994 estimated at £31.1 billion. On 1 January 1993 the Government deficit, the National Debt, stood at just over £200 billion. Of this, £16.5 billion were paid in debt interest in 1992-3. Britain's contribution to the European Community for the fiscal year 1992-93 was £2.46 billion
© Martin Gilbert 1993